# Green Energy and Technology

Lucas Reijnders • Mark A.J. Huijbregts

# Biofuels for Road Transport

## A Seed to Wheel Perspective

Lucas Reijnders, PhD
Institute for Biodiversity and Ecosystem
Dynamics (IBED)
Faculty of Science
University of Amsterdam
Nieuwe Achtergracht 166
1018 WV Amsterdam
The Netherlands

Mark A.J. Huijbregts, PhD
Department of Environmental Science
Faculty of Science
Radboud University Nijmegen
6500 GL Nijmegen
The Netherlands

ISBN 978-1-84996-823-2 e-ISBN 978-1-84882-138-5

DOI 10.1007/978-1-84882-138-5

Green Energy and Technology ISSN 1865-3529

A catalogue record for this book is available from the British Library



*Cover design*: WMXDesign, Heidelberg, Germany

Printed on acid-free paper

9 8 7 6 5 4 3 2 1

springer.com

# Contents

# Chapter 1
# Transport Biofuels: Their Characteristics, Production and Costs

## 1.1 Introduction

Production and use of transport biofuels have a history of considerable length. The prototype of the Otto motor, which currently powers gasoline cars, was developed for burning ethanol and sponsored by a sugar factory. The Ford Model T (Tin Lizzy) did run on ethanol. In the early twentieth century, ethanol-fuelled cars were praised because they experienced less wear and tear, were quieter and produced a less smoky exhaust than gasoline-fuelled cars (Dimitri and Effland 2007). Also in the early twentieth century, a significant part of train locomotives in Germany were powered by ethanol (Antoni et al. 2007). In the same country, ethanol from potato starch was used in gasoline as an anti-knocking additive between 1925 and 1945 (Antoni et al. 2007). In the 1930s, ethanol produced from starch or sugar made something of a comeback as road transport fuel in the Midwestern states of the USA, because agricultural prices were very depressed (Solomon et al. 2007). Also in the 1930s, the Brazilian government stimulated gasoline blends with 5% bioethanol.

Early demonstrations of the diesel motor around 1900 in Paris and St Petersburg were with a variety of plant- and animal-derived oils. These were thought especially interesting for use in tropical and subtropical countries, where the relatively high viscosity of such oils, if compared with fossil diesel, is less of a problem than in colder countries (Knothe 2001). The first patent on making fatty acid esters (biodiesel) was awarded in 1937 and applied in 1938 to powering buses in Belgium (Knothe 2001). During the Second World War, vegetable oils re-emerged as fuels for diesel motors in countries like Brazil, Argentina and China (Knothe 2001). In Japan, soybean oil was used to power ships, pine root oil was used as a high-octane motor fuel and biogenic butanol was used in airplanes (Tsutsui 2003). The Japanese navy conducted extensive research on the production of diesel fuel from coconut oil, birch bark, orange peel and pine needles (Tsutsui 2003). Also, during the Second World War, substitutes for mineral-oil-based gasoline and kerosene were produced in China by the catalytic cracking of vegetable oils (Knothe 2001). Furthermore, thermal destruction of wood was used for producing road transport fuel during the World Wars in Europe (Reed and Lerner 1973).

The post-World War II re-emergence of transport biofuel use dates from the 1973 hike in petroleum prices, or the ‘first oil crisis’. Tax reductions, subsidies, support for research and development, obligations to fuel providers and artificially high fuel economy ratings for flex fuel cars, which are suitable for high percentages of biofuel in transport fuel, were important government instruments used in this re-emergence (Demirbaş 2007; Szklo et al. 2007; Tyner 2007; Wiesenthal et al. 2008). By now, large sums of money are involved in such support. It has been estimated that in 2006, about US $11 billion was spent on public support measures by the USA, Canada and the European Union (OECD 2008).

Due to the first oil crisis of 1973, Brazil decided to reduce its dependence on the import of mineral oil by establishing a National Alcohol Program to supply vehicles. This program started in 1975, using sugar cane as a feedstock. A second program stimulating the use of ethanol began in the USA in 1978, using mainly corn and to a much lesser extent sorghum as feedstocks (Wheals et al. 1999; Wang et al. 2008a). In the USA, arguments for subsidizing the production of bioethanol since 1978 have included energy security, supporting farm prices and incomes and improvement of air quality (Tyner 2007). Several Canadian provinces started out using 5–10% ethanol–gasoline mixtures in the 1980s (Szklo et al. 2007). The ‘rediscovery’ of biodiesel occurred in the 1980s. Biodiesel initiatives were announced in 1981 in South Africa and in 1982 in Germany, New Zealand and Austria (Körbitz 1999). In Europe, substantial production of biodiesel started from about 1987 and in the USA from the 1990s (Knothe 2001). The relatively large production of biofuels in countries such as Germany, France, Italy, Austria and Spain had much to do with an interest in the development of new agricultural markets (Di Lucia and Nilsson 2007). Geopolitical worries about the supply of crude mineral oil and price rises affecting this dominating feedstock for current transport fuels furthered a rapid increase in biofuel production in the twenty-first century, especially after 2004 (Heiman and Solomon 2007).

The production of conventional mineral oil is likely to peak in the coming decades (GAO 2007; Bentley et al. 2007; Kaufmann and Shiers 2008). An adequate supply thereof may therefore become increasingly expensive and difficult. This has led to calls to – in the words of former US president G. W. Bush – kick the oil ‘addiction’ (Bush 2006). Timeliness of a transition to alternative fuels has been stressed (Kaufmann and Shiers 2008). ‘Home-grown’ biofuels, especially, have been argued to be suitable for energy security (Tyner 2007). There is also much concern about the pollution originating in the burning of fossil fuels. Recently, the effects thereof on climate have become important on the international political agenda. This, in turn, has led to increasing calls to reduce the emission of greenhouse gases, such as $CO_2$. Such calls extend to transportation because worldwide transport accounts for about 22% of the total use of primary energy and is overwhelmingly mineral oil based (de la Rue du Can and Price 2008). For instance, regarding the USA, mineral oil accounted in 2006 for about 97.8% of total transport energy use (Heiman and Solomon 2007). Worldwide, the consumption of petroleum products represents 94% of energy use in the transportation sector (de la Rue du Can and Price 2008), whereas in 2004, about 60% of all mineral oil was used for transportation (Quadrelli and Peterson 2007). Proponents of biofuels have argued that replacement of mineral oil by biofuels is a good way to reduce greenhouse gas emissions.

It has furthermore been stated that the potential for replacing fossil transport fuels with biofuels is very substantial indeed. de Vries et al. (2007) have suggested that by 2050, up to 300 EJ ($= 300 \times 10^{18}$ J) of liquid biofuels may be produced worldwide. An even higher estimate for liquid biofuel production by 2050 (455 EJ) has been proposed by Moreira (2006). Such amounts can in all probability cover demand for transport fuels in 2050, as the 2007 primary energy consumption for transport amounted to about 100 EJ (de la Rue du Can and Price 2008). Use of transport fuels by means of transport ('end use') was probably in the 85–90 EJ range, with the remaining amount used for winning, refining and distribution (Colella et al. 2005; EUCAR et al. 2007; Winebrake et al. 2007). The potential importance of biofuels in replacing fossil transport fuels is by now much stressed by the Brazilian government. In Brazil, ethanol from sugar cane is currently a substantial transport biofuel. In 2004, its share in energy for road transport was near 14% and in 2007 about 20% (OECD 2008). In 2006, 70% of the new cars sold in Brazil were 'flex cars', able to run on either 100% ethanol or a fossil fuel–ethanol blend (Quadrelli and Peterson 2007). The claims about the benefits and potential of transport biofuels have, however, been contested. And the resulting debate has been much fuelled by the high food prices in 2008, which have been partially linked to increasing transport biofuel production (OECD-FAO 2007).

This book will give a seed-to-wheel perspective on biofuels for road transport and will deal with a number of environmental issues that have emerged in the current biofuel debate. This first chapter is introductory and structured as follows: firstly, Sects. 1.2–1.6 will deal with the physical basis and the variety of biofuels and the ways to produce and apply them in transport. Secondly, in Sect. 1.7, developments in production volume, costs and prices will be discussed. Thereafter, in Sect. 1.8, the debate on the pros and cons of transport biofuels that has emerged will be briefly surveyed, and the rest of the book will be outlined.

## 1.2 The Physical Basis for Biofuels

Biofuels are ultimately based on the ability of photosynthetic organisms to use solar irradiation for the conversion of $CO_2$ into glucose ($C_6H_{12}O_6$) and subsequently into biomass; the overall reaction for the conversion into glucose usually being:

$$6\,CO_2 + 6\,H_2O \rightarrow C_6H_{12}O_6 + 6\,O_2$$

Some photosynthetic bacteria may not produce oxygen but give off elemental sulphur.

In practice, only part of incident solar radiation is captured by plants. And of the solar irradiation captured by plants, only a part (approximately 43–45% of radiation in the visible part of the spectrum for land plants) is photosynthetically active (Sinclair and Muchow 1999; Vasudevan and Briggs 2008).

The synthesis of glucose is powered by light reactions generating NADPH, ATP and $O_2$. Thereafter, the reactions can proceed in the dark. In these reactions, collectively known as the Calvin cycle, ATP, NADPH and $CO_2$ are converted into glucose, $NADP^+$, ADP and phosphate.

The first enzyme of the Calvin cycle is ribulose bisphosphate carboxylase. As ribulose bisphosphate carboxylase is sensitive to oxygen, photorespiration is important to protect the enzyme. When $CO_2$ levels in the atmosphere increase, protection by photorespiration can be reduced. At the present atmospheric concentration of $CO_2$, in most plants, photorespiration leads to the release of up to about 50% of the $CO_2$ originally fixed by photosynthesis. These plants are called $C_3$ plants. This name is linked to the first product of photosynthesis that contains 3 C atoms, 3-phosphoglyceric acid. All large trees are $C_3$ plants (Heaton et al. 2008). More recently in the evolution of terrestrial plants, a retrofit to the Calvin cycle has emerged that reduces the need for photorespiration. The plants having such a retrofit are called $C_4$ plants. This name is again linked to the first product(s) of photosynthesis that are organic acids with 4 C atoms. Examples of $C_3$ terrestrial plants relevant to biofuels are wheat, rapeseed, soybean, sunflower, eucalyptus, sugar beet, potato, poplar, coconut, cassava, cotton and *Jatropha*, while examples of $C_4$ plants are sugar cane, corn (maize), switchgrass, sorghum, millet, and *Miscanthus*.

Natural $C_4$ species tend to be better adapted to relatively warm climates than $C_3$ species. However, breeding and selection have changed the temperature response in a number of $C_3$ and $C_4$ species. Thus, there are now $C_3$ species that do optimally in relatively warm climates (e.g. cotton) and $C_4$ species, such as corn varieties which have been well adapted to temperate climates (El Bassam 1998). The reduced need for photorespiration in $C_4$ species is reflected in a higher maximum theoretical efficiency in the conversion of solar irradiation into biomass.

For $C_4$ plants on land at the present concentration of $CO_2$, the maximum theoretical efficiency is estimated at 5.5–6.7% and for $C_3$ plants on land at 3.3–4.6% (Hall 1982; El Bassam 1998; Kheshgi et al. 2000; Heaton et al. 2008). For algae, a theoretical efficiency varying between 5.5 and 11.6% has been suggested (Heaton et al. 2008; Vasudevan and Briggs 2008). Actual efficiencies in commercial cultivation are much lower, as will be discussed in Chap. 2. Most transport biofuels are derived from photosynthetic organisms, though there is also a limited supply of biofuels derived from animals (based on, for example, yellow grease and animal meal).

## 1.3 Biofuel Varieties

There are a variety of ways to use biofuels for transport. The first category focuses on electric traction, which currently accounts for about 1% of energy use in the transportation sector worldwide (de la Rue du Can and Price 2008). Electric traction is common in train transport, but there are also ships powered by electricity, and a battery-powered small airplane has been demonstrated (Sanderson 2008). All-electric cars currently have limited application, but more recently there has been a rapid increase in the use of hybrid cars that use both internal combustion engines and electromotors (Mom and Kirsch 2001; Würster and Zittel 2007; Høyer 2008).

Electricity can, for instance, be generated in power plants fired by biomass and stored in batteries . Also, electricity can be generated by onboard fuel cells fed with, for example, $H_2$ derived from biomass or $H_2$-producing organisms. Hydrogen used in fuel cells is, from a life cycle perspective, more energy efficient than the application of $H_2$ in Otto or diesel motors (EUCAR et al. 2007; Hussain et al. 2007; Kleiner 2007). Fuel cells may also be used for the propulsion of ships and airplanes (Littlefield and Nickens 2005; Lapeña-Rey et al. 2008; Sanderson 2008). Introduction of hydrogen as a major transport fuel requires concerted action of many stakeholders (Würster and Zittel 2007) and includes large changes in fuelling infrastructure and a major effort to reduce fire and explosion risks (MacLean and Lave 2003; Agnolucci 2007; Astbury and Hawksworth 2007; Markert et al. 2007; Melaina 2007; Ng and Lee 2008). Also, major advances in several key components of motorcars are necessary for a successful large-scale introduction of all-electric or $H_2$-powered cars (Chalk and Miller 2006; Matheys et al. 2007; Høyer 2008; Lache et al. 2008; Samaras and Meisterling 2008).

In practice, wood, animal wastes, harvest residues, municipal and industrial organic wastes, landfill gas, 'energy' grasses (such as reed canary grass) and vegetable oils have been used in power generation (e.g. Reijnders and Huijbregts 2005; Berggren et al. 2008; Heinimö 2008; Junginger et al. 2008; Reijnders and Huijbregts 2008). Sewage sludges and wastewater treatment sludges are also applied, though these tend to be net users instead of net producers of energy due to their high water content (Wang et al. 2008c).

There is, furthermore, scope for the co-production of electricity and ethanol from sugar cane (Macedo et al. 2008). In producing electricity, both direct burning of biomass and burning after gasification or fermentation are practiced (Wheals et al. 1999). Problems in generating electricity from biomass have arisen due to slagging, corrosion and fouling mainly linked to the presence of inorganic elements such as Cl and K; in the case of gasification, fouling has also been linked to tar formation (Monti et al. 2008). Ways to decrease such problems, such as lowering Cl and K concentrations by judicious choice of feedstocks, have been researched (Monti et al. 2008), though there are types of biomass, such as macroalgae, that still appear unsuitable for direct combustion or gasification (Ros et al. 2009).

The second possibility is to produce liquid or gaseous biofuels that can be burnt in transport engines that currently burn fossil fuels. In 2006, such biofuels accounted for about 1% of energy use in the transportation sector worldwide (de la Rue du Can and Price 2008). Various engines operate under a variety of conditions, and not all liquid and gaseous biofuels are suitable to all applications. Quantitatively speaking, two engine types dominate road transport, and also transport in general: the diesel engine and the Otto motor. A variety of gaseous and liquid biofuels produced have been proposed for these engine types. As to the way these biofuels are produced, most of them can be allocated to three categories (Ahman and Nilsson 2008). The first category relies on the biochemical conversion of biomass into transport biofuels. Biochemical conversion is now used for the production of ethanol, butanol and methane. The second category is based on lipids (oils and fats) derived from organisms. Such oils may be applied directly or after processing (e.g. transesterification or catalytic cracking). The third category uses thermochemical conversion of biomass via pyrolysis or gasification into a variety of fuels.

A part of the transport biofuels which have been proposed are currently produced on an industrial scale and widely applied in means of transport. Ethanol obtained from starch or sugar by fermentation and biodiesel based on lipids from terrestrial plants are currently the main transport biofuels. Other substances that have potential as transport biofuels are produced on an industrial scale but hardly or not applied in Otto and diesel motors. A third category of transport biofuels include those in the laboratory and pilot plant stage. All these are shown in Table 1.1.

**Table 1.1** Production and application of a variety of transport biofuels

| Industrial-scale production and applied in Otto and diesel motors | Production | Application |
|---|---|---|
| Ethanol | By fermentation from starch or sucrose | Mostly in Otto motors, pure or as blend |
| ETBE (*tert*-butylether of ethanol) | Ethanol produced by fermentation from starch or sucrose | In Otto engines, as blend |
| Biodiesel (ethyl- or more often methylester from long chain fatty acids) | Fatty acid ester from biogenic lipids by transesterification | In diesel motors, pure or as blend |
| **Industrial-scale production, but hardly applied in Otto or diesel motors** | **Production** | **Application** |
| Methane | By anaerobic conversion from a wide variety of biomass types | Combined use with gasoline or diesel in Otto or diesel engines |
| Vegetable lipids (oils), e.g. palm oil, coconut oil | Extraction from oil crops | Currently limited application in diesel motors |
| Turpentine | Co-product from wood processing (e.g. paper production) | May be mixed into gasoline and diesel (Yumrutaş et al. 2008) |
| Ethanol | By fermentation from wood hydrolysate containing sugars | Mostly in Otto motors |

**Table 1.1** (continued)

| Production at the pilot plant or laboratory stage | Production | Application |
|---|---|---|
| Methanol, also as MTBE (*t*-butylester of methanol) | Via synthesis gas from glycerol or biomass; microbially from sugar beet pulp (Antoni et al. 2007) | In Otto motors; methanol may also be used in fuel cell cars, though relative activity of methanol in fuel cells is much lower than of $H_2$ (Lewis 1966) |
| Dimethylether (DME) | Via synthesis gas from biomass by gasification with pure oxygen (Arcoumanis et al. 2008) | Proposed as alternative to diesel in diesel engines; also suitable for gas turbines |
| Butanol, also as BTBE (*t*-butylester of butanol) | Butanol by fermentation from sugar/starch or (hemi)cellulose | In Otto motors, turbofan engines |
| Biohydrogen | By photosynthetic algae, via fermentation by $H_2$-producing microbes, by photo-induced reforming or via synthesis gas | In fuel cells or engines |
| Hydrocarbons | Via synthesis gas from biomass or components/ conversion products thereof, by cracking/deoxygenation of lipids or cracking of microalgal hydrocarbons | In Otto and diesel motors |

The energy contents of the liquid and gaseous transport biofuels mentioned in Table 1.1 may be different from the fossil petrol and diesel that they replace. Table 1.2 gives a survey of the energy contents (lower and higher heating values) in megajoules (MJ) of the liquid fossil and biofuels per kilogram (kg) and per litre (l). The lower heating value (LHV) represents net energy content, and the higher heating value (HHV) represents gross energy content (including the heat of condensation of water vapour produced by combustion (Piringer and Steinberg 2006)).

The differences in heating values indicate that when the amount of transport kilometres for a full tank is to be maintained, a substantial adaptation of tank size may be necessary when transport fuels contain high percentages of biofuels with relatively low heating values, such as dimethylether and ethanol (Semelsberger et al. 2006). This is not the only adaptation that may be necessary when switching to biofuels. Table 1.3 gives a brief summary of other adaptations for a number of biofuels.

**Table 1.2** Energy content (lower and higher heating values, with the latter including the latent heat of vaporization) for liquid transport fossil and biofuels per kilogram and litre (Anonymous 2006; Hammerschlag 2006; European Union 2008; Savage et al. 2008)

| Transport fuel | Lower heating value by weight ($MJ\,kg^{-1}$) | Lower heating value by volume ($MJ\,l^{-1}$); for liquid biofuels only | Higher heating value by weight ($MJ\,kg^{-1}$) |
|---|---|---|---|
| Ethanol | 26.4 | 21.2 | 29.8 |
| ETBE | 36.0 | 26.7 | 39.2 |
| Biodiesel (average for fatty acid methylesters) | 37.3 | 32.8 | 40.2 |
| Methanol | 19.8 | 15.6 | 22.9 |
| MTBE | 35.2 | 26.0 | 38.0 |
| Dimethylether | 28.4 | 20.3 | 31.7 |
| Butanol | 35.4 | 27.8 | |
| Palm oil | 37.0 | 34.9 | |
| Fischer–Tropsch diesel made from natural gas | 44.0 | 34.3 | 45.5 |
| Methane | 50.0 | – | 55.2 |
| Diesel (from mineral oil, European) | 41.2 | 35.7 | 45.6 |
| Gasoline (also called petrol) (from mineral oil, European) | 42.7 | 31.0 | 46.5 |
| Hydrogen | 120 | – | 141.8 |

**Table 1.3** Problems and adaptations necessary for the use of biofuels

| Biofuel | Problems and adaptations |
|---|---|
| Ethanol | – Ethanol is relatively corrosive, and ethanol–gasoline blends may separate in pipelines; this limits the scope for pipeline transport. Also, ethanol is hygroscopic, and high water concentrations may lead to phase separation. So, in storage and distribution, exposure to water should be severely limited (Antoni et al. 2007; Atsumi et al. 2008).<br>– Limited admixture of ethanol (whether or not as ETBE: the tertiary butylether of ethanol) up to 5% is possible without adaptation of cars. If ethanol–fossil hydrocarbon blends with percentages of ethanol over 5% are used, however, changes in cars are needed (Antoni et al. 2007). Such changes regard the fuel-sending unit, the fuel injector, the fuel filter, fuel management and flame arrestors. When the percentage of bioethanol becomes 85 or 100%, changes necessary for the engine become substantial (Antoni et al. 2007; Hammond et al. 2008). This has led to the development of flex vehicles that are able to run on blends with high percentages of ethanol, and also on conventional petrol. |

**Table 1.3** (continued)

| Biofuel | Problems and adaptations |
|---|---|
| Vegetable oil | – High viscosity may give rise to increased fuel consumption, to increased emissions of CO and hydrocarbons and to engine durability problems (Agarwal and Agarwal 2007; Scholz and da Silva 2008).<br>– Oils with unsaturated fatty acids may be subject to oxidative instability (Vasudevan and Briggs 2007). Such instability may be corrected by hydrogenation (Mikkonen 2008). However, saturated fatty acids are more prone to form crystals at relatively low temperatures, and thus their presence is also subject to limitation.<br>– To the extent that vegetable oils are suitable, use thereof is associated with substantially increased maintenance (Cloin 2007).<br>– In aircraft, vegetable oils freeze at normal cruising temperatures and have relatively poor high temperature thermal stability characteristics in the engine (Daggett et al. 2007). |
| Fatty acid esters (biodiesel) | – At low fuel temperature, viscosity of biodiesel and precipitate formation may still become unacceptable (Kerschbaum et al. 2008). Unacceptable viscosity may be associated with piston ring sticking and severe engine deposits (Kegl 2008). Also, at low temperatures, there may be more cold-starting problems (Hammond et al. 2008).<br>– Saturated fatty-acid-based biodiesel is relatively prone to crystal formation at low temperatures, more so when the carbon chains are longer. Ozonization, lowering the content of saturated fatty acids and the use of fatty acids with shorter carbon chains have been proposed as ways to 'winterize' biodiesel (Kerschbaum et al. 2008; Ramos et al. 2008).<br>– Precipitate formation at low temperatures may also be linked to the presence of (plant-derived) steryl glucosides (Tang et al. 2008).<br>– When a substantial percentage of biodiesel is present in the transport fuel, especially in older cars, there may be a need to change fuel hoses and seals, because these will otherwise corrode (Radich 2007; Hammond 2008).<br>– The amount of free alcohol in biodiesel should be kept very low to prevent accelerated deterioration of rubber seals and gaskets (Abdullah et al. 2007).<br>– The solvent property of biodiesel may be conducive to loosening deposits in fuel systems, which may lead to clogging of fuel lines and filters and, more in general, there may be a need for more frequent oil and fuel filter changes when biodiesel is used (Radich 2007; Hammond et al. 2008).<br>– In aircraft, only the admixture of low percentages of biodiesel in jet fuel is acceptable to prevent freezing (Wardle 2003).<br>– Storage of biodiesel should be such that oxidative and hydrolytic deterioration are prevented. Similarly, the presence of water should be prevented, as this is conducive to the growth of micro-organisms (Abdullah et al. 2007). |
| Methane | – Supply system has to be adapted to store and handle methane.<br>– Cars have to be adapted to dual fuelling (Björesson and Mattiasson 2008).<br>– Optimum use of methane requires engine modifications (Björesson and Mattiasson 2008; Hammond et al. 2008). |

**Table 1.3** (continued)

| Biofuel | Problems and adaptations |
|---|---|
| Dimethylether | – New storage and fuel delivery systems are needed (Semelsberger et al. 2006).<br>– Provisions have to be made to reduce leakage in pumps and fuel injectors (Semelsberger et al. 2006).<br>– Adaptation of engines or the use of additives to solve problems with lubricity is necessary (Semelsberger et al. 2006; Arcoumanis et al. 2008).<br>– Modifications of engines are needed to prevent corrosion (Arcoumanis et al. 2008). |

## 1.4 Virtual Biofuels

There is also an option which may be called *virtual* biofuels. The pyrolytic production of 'black carbon' (also charcoal, biochar) from biomass has been advocated as an alternative to biofuels (Fowles 2007). Such black carbon would be added to soils, where it said to be 'very stable' and able to fulfil useful functions. This is argued to offset $CO_2$ emissions (Lehmann et al. 2006; Fowles 2007; Mathews 2008a). Saunders (2008) has proposed to landfill purpose-grown biomass as a 'virtual biofuel', which he considers 'more practical, economic and immediate' than the use of actual biofuels from lignocellulosics. There are also 'climate compensation schemes' offered to users of transport (especially car and air transport) to offset their emissions of $CO_2$. Planting trees tends to be major contributor to such schemes. The idea behind this is that the C emitted as $CO_2$ into the atmosphere due to the burning of fossil fuels will be sequestered in biomass. For instance, it has been estimated that reforestation of abandoned tropical land may lead to an aboveground C sequestration of approximately 1.4 Mg $ha^{-1}$ $year^{-1}$ and a sequestration in soils of approximately 0.4 Mg C $ha^{-1}$ $year^{-1}$ over an 80–100-year period (Silver et al. 2000). Offsetting by forest conservation or reforestation leads to much lower costs for the alleged reduction of $CO_2$ emissions from transport than the production of biofuels.

The obvious question about these proposals is: are virtual biofuels indeed a solution to the impact of fossil transport fuels on climate? This depends on the duration of carbon sequestration in virtual biofuels. Before being used as transport fuel, mineral oil was destined to remain for many millions of years outside the carbon cycle in which biomass participates, and one may argue that to really offset the use thereof, C in black carbon, purpose-grown landfilled biomass and forests should also remain outside the biogeochemical carbon cycle for many millions of years or 'in perpetuity'. Also, one may focus on $CO_2$ emitted due to the consumption of fossil fuels. Full elimination of the effect of the $CO_2$ emission from fossil transport fuels on the climate is expected to take a very long time. One quarter of fossil-fuel-derived $CO_2$ remains airborne for several centuries, and complete removal may take 30,000–35,000 years (Archer 2005; Hansen et al. 2008). So it may be argued that the sequestration in biochar, landfills or forests should at least be for many thousands of years.

Whether C sequestration for at least many thousands of years applies to biochar has been studied in a limited way (Eckmeier et al. 2007; Wardle et al. 2008). There have been reports about the decomposition of biochar and oxidation of the aromatic backbone of biochar, partly depending on the production procedure (Lehmann et al. 2006; Steiner et al. 2007). However, there is also evidence that carbon black particles may persist in soils over thousands of years (e.g. Carcaillet and Talon 2001; Long et al. 2007), allowing for the possibility that part of buried biochar sequesters C for a very long time indeed (Lehmann et al. 2006). Secondly, there is evidence that biochar may have an effect on soil biological processes: experimental data suggest that this effect may result in loss of native soil carbon (Wardle et al. 2008). As it stands, it would seem likely that the net carbon sequestration by biochar is partial and may show a decrease over time.

Landfilled purpose-grown biomass will not for thousands of years or in perpetuity remain outside the biogeochemical carbon cycle. Landfilled biomass will largely be converted into $CH_4$ by anaerobic processes. This has the added disadvantage that $CH_4$ is, over a 100-year period, 21 times a more potent greenhouse gas than $CO_2$, the main carbonaceous product of biofuel combustion (Barlaz 2006). The rate of conversion of biomass into methane is dependent on a variety of factors, including temperature, lignin content, moisture and pH, and such conversion is often a matter of decades (Barlaz 2006, Themelis and Ulloa 2007). Capture of $CH_4$ emitted by landfills is an option but has in practice limited efficiency (Themelis and Ulloa 2007).

There is also a major snag with planting forests as virtual biofuels. Storage of carbon in trees should be guaranteed for many thousands of years. At the level of individual trees, this is impossible, as storage in perpetuity or for many thousands of years is well beyond the maximum lifespan of tree species. And guarantees at the level of forests are also a problem. Current ‘climate compensation schemes’ have guarantees for forests that do not exceed a hundred years. Even this guaranteed timeframe is questionable in view of (increasing) risks that forests may be destroyed by wildfires and extreme weather events such as storms and droughts (Kirilenko and Sedjo 2007; Gough et al. 2008; Nepstad et al. 2008). The social arrangements safeguarding forests are also unlikely to persist for many thousands of years or in perpetuity. So it would seem that forests as virtual biofuels rather delay than fully offset the emission of $CO_2$ from fossil transport fuels. There is also another problem with planting forests to limit the increase in atmospheric $CO_2$. This has been called ‘leakage’ (Sathaye and Andrasko 2007; Ewers and Rodrigues 2008). When forestry projects are established, people dependent on that area may move elsewhere, where they may reduce C stocks. There has been a series of case studies regarding this phenomenon. In some cases, high levels of leakage have been demonstrated. For instance, Boer et al. (2007) studied forestation projects in the Jambi province of Indonesia and found that reductions in C stock due to leakage exceeded gains in C stock linked to forestation over a 10-year period. Other forestation projects showed lower leakage, and worldwide, an average percentage of about 50% leakage seems to be associated with forestation projects (Sathaye and Andrasko 2007). So forestry as a virtual biofuel is subject to major problems when it comes to full compensation of fossil fuel consumption.

Apart from problems with C sequestration over at least thousands of years, a main problem is that virtual biofuels clearly would not solve the problem of dependence on mineral oil, as one cannot drive or fly on virtual fuels. For this reason, the 'real' biofuels that are used or have been proposed for use as transport biofuels will be the main topic of this book.

## 1.5 Production and Application of Liquid and Gaseous Transport Biofuels

In this book, the focus will be largely on liquid and gaseous transport biofuels that can replace fossil transport fuels. In this section, a brief survey will be given of the ways to produce liquid and gaseous transport biofuels from a variety of feedstocks. These are summarized in Fig. 1.1.

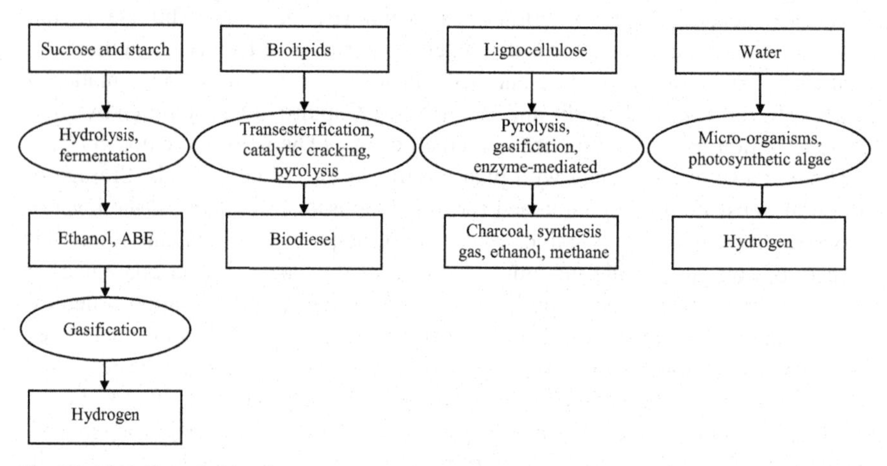


**Fig. 1.1** Biofuel production steps

### *1.5.1 Biofuels from Terrestrial Plants and Animals*

#### Sucrose- or Starch-Based Biofuels

Sucrose (from sugar cane, sugar beet and sweet sorghum) and starch (from starch crops such as corn, grain sorghum, potato, Jerusalem artichoke, cassava, rye, barley, sago palm and wheat) can be converted into ethanol by hydrolysis and fermentation. There is also the possibility of converting sugar in whey and starch and sugar in wastes (potato peel, spoiled fruit) into ethanol by fermentation (Acharya and Young 2008). The fermentation used to produce ethanol is usually yeast based. The reaction starts with a $C_6$ sugar and is:

$$C_6H_{12}O_6 \rightarrow 2CO_2 + 2C_2H_5OH \text{ (ethanol)} .$$

Ethanol has its disadvantages vis-à-vis fossil-based gasoline. Its lower heating value is considerably lower (see Table 1.2), and it is hygroscopic and more corrosive (cf. Table 1.3). Against this background there have been proposals to convert ethanol to hydrocarbons ('biogasoline') (Tsuchida et al. 2008) or $H_2$ (Ni et al. 2007; Kondarides et al. 2008). However, in practice, ethanol may do well as transport fuel in Otto motors, both as such and as a mixture with fossil hydrocarbons (Szklo et al. 2007). Ethanol may also be applied in a mixture with fossil diesel fuel in diesel motors – with an additive to prevent phase separation (Fernando and Hanna 2004; Antoni et al. 2007; Wang et al. 2007; Song et al. 2007). In Otto motors, ethanol is used as a gasoline extender, octane booster and oxygenate suitable for driving during winter in temperate climates (MacLean and Lave 2003). Claims have been made that admixture of ethanol improves the fuel efficiency of Otto motors, but available evidence (Roberts 2008; Kamimura and Sauer 2008) suggests that differences in average fuel efficiency are not statistically different from the differences in heating value. Ethanol can also be used to produce ETBE (ethylester of *t*-butanol) or ethylesters of fatty acids, which can be applied in Otto and diesel motors respectively. Currently the production thereof is in chemical reactors. For the combined production of ethanol from sugars and ethylesters of fatty acids, a synthesis employing genetically modified *Escherichia coli* has been demonstrated (Kalscheuer et al. 2006).

Starch and sucrose may also serve as a basis for fermentation into butanol, or to be more precise, a mixture of acetone, butanol and ethanol (ABE). After World War II, bacterial fermentation generating ABE from starch and sucrose was applied on an industrial scale in a wide variety of countries. This production process ultimately succumbed to the price competition of petrochemical butanol (Ng et al. 1983; Reinharz 1985; Jones and Woods 1986; Gutierrez et al. 1998; Zverlov et al. 2006; Chiao and Sun 2007; Ezeji et al. 2007; Qureshi et al. 2008a). There currently is pilot-scale industrial production of butanol by bacterial fermentation processes starting with starch or sucrose and a substantial amount of research and development aimed at 'engineering out' the production of acetone and ethanol (Wackett 2008). Up to about 18% ABE may be mixed with fossil diesel fuel, which is then suitable for powering diesel motors (Willke and Vorlop 2004). Butanol can be mixed into gasoline for use in Otto motors, as such or after esterification with *t*-butanol (Scott and Bryner 2006; Antoni et al. 2007; Ezeji et al. 2007). Butanol is a biofuel that can also be used in high thrust-to-weight applications such as aircraft engines. Butanol has the added advantages that, unlike ethanol, it will not solidify at the low temperatures of high altitudes at which airplanes operate and that it is not hygroscopic. Disadvantages are that the concentration of butanol achievable by fermentation is currently low and that the boiling point is high, which necessitates relatively high energy inputs for butanol distillation (Fortman et al. 2008; Hayes 2008).

It has also been shown that the production of branched chain butanols (isobutanol, 2-methyl-1-butanol, 3-methyl-1-butanol) from glucose is possible using metabolic engineering of micro-organisms (Atsumi et al. 2008). Such branched chain butanols can also be mixed into gasoline.

### Lipid-Based Biofuels from Terrestrial Plants or Animals

A wide variety of plants produce lipids that can be used as a basis for transport fuels. Main suppliers of lipids are currently plants producing edible oils such as rapeseed or canola, sunflower, oil palm, coconut and soybean. There is also limited use of non-edible lipids such as *Jatropha* oil and of animal fats ('yellow grease') and very limited use of algal lipids. A number of additional potential vegetable sources of lipids have been suggested (Sims et al. 2006; Shao and Chu 2008), and lipids have also been produced microbially from sugars (Zhou et al. 2008). The use of edible oils often has the advantage that co-products may be used in animal feed production. This may be different for non-edible oils. For instance, the oil cake of many of the current *Jatropha* varieties is not suitable for feeding livestock because of the presence of toxic compounds such as phorbolester and curcin (Carvalho et al. 2008; Sujatha et al. 2008), but such toxic *Jatropha* oil cake can be anaerobically converted to methane (Achten et al. 2009) or burned to supply energy.

The lipids used for biofuel production mainly consist of triacylglycerol, in which the acyl groups are fatty acids (Agarwal and Agarwal 2007). In principle, a variety of lipids can be burnt as such in diesel motors, more of them in warm climates. Vegetable oils are used to a limited extent as transport fuel. For instance, there is a significant use of coconut oil in motorcars on the Pacific islands (Cloin 2007), and in Europe there is limited use of rapeseed oil in heavy-duty vehicles. However, for most applications, viscosity is too high. This problem may be solved by dilution, microemulsification and transesterification (Canakci and Sanli 2008). In practice, the solution is mostly transesterification to produce biodiesel, which is compatible with fossil diesel fuel and leads to a more limited increase in maintenance costs. In transesterification, the glycerol OH groups are replaced by the OH groups of either ethanol or, more commonly, methanol.

Transesterification to produce biodiesel from lipids can proceed with the help of an inorganic base catalyst (e.g. NaOH, KOH or $NaOCH_3$). This approach is widely applied in commercial biodiesel production (Canakci and Sanli 2008). Potential alternatives are the use of insoluble inorganic catalysts (Shu et al. 2007; Li et al. 2007; Vasudevan and Briggs 2008) and the use of an enzyme: lipase (Harding et al. 2008). These alternatives are under development (Abdullah et al. 2007; Ranganathan et al. 2008). Transesterification by superheated or supercritical alcohols (that are not sensitive to free fatty acids and water) has also been studied (Marchetti et al. 2007; Joelianingsih et al. 2008). In the case of lipids that are characterized by the presence of greater than 0.5–1% free fatty acids (that react with base catalysts to soap) – often waste lipids – the use of both homogeneous and heterogeneous acid catalyzed transesterification has been advocated (Abdullah et al. 2007; Vasudevan and Briggs 2008; You et al. 2008; Canakci and Sanli 2008; Park et al. 2008). Alternatively, free fatty acid levels can be reduced to less than 0.5% by the use of ion-exchange resins (Özbay et al. 2008) or the admixture of virgin lipids.

An alternative option to transesterification is to catalytically remove oxygen from the triacylglycerol, while adding hydrogen. This gives rise to the synthesis of propane and mixtures of hydrocarbons (paraffins) that have diesel-like properties and can also be used in kerosene blends (Holmgren et al. 2007). Such a deoxygenation process is currently commercially exploited (Rantanen et al. 2005). Still another way of converting virgin or used vegetable oil in transport fuels uses catalytic cracking or pyrolysis (heating in the absence of oxygen). In the latter case, this has to be followed by upgrading of the bio-oil that is a product of pyrolysis. In this way, one may produce fuels that are suitable for application in diesel or Otto motors or as a substitute for kerosene that is applicable in air transport (Milne et al. 1990; Knothe 2001; Demirbaş and Kara 2006; Dupain et al. 2007; Ooi and Bhatia 2007; Tamunaidu and Bhatia 2007). It has been noted that in the case of cracking unsaturated lipids, the product may contain relatively large amounts of aromatics (Dupain et al. 2007). Also, alkane synthesis from lipids by the bacterium *Vibrio furnissii* has been reported (Fortman et al. 2008). Finally, there are efforts to produce fatty acid ethylesters and hydrocarbons by re-engineering metabolism in heterotrophic microorganisms (Wackett 2008). Ultimately, this way of producing biofuels is critically dependent on cheap sources of carbohydrates (Rotman 2008).

In Germany, a major producer of biodiesel, glycerol that is generated by transesterification of oils and fats is anaerobically converted into methane (see below). Glycerol can also be gasified to synthesis gas (mainly CO and $H_2$). Synthesis gas (also called 'syngas') may be converted into methanol. Methanol, in turn, can be mixed into gasoline up to 15% by volume and applied in Otto motors without major adaptations. Methanol can also be used for the production of biodiesel, as an admixture in diesel (Cheng et al. 2008) and to produce methyl *tert*-butylether (MTBE) to be applied in petrol for use in Otto motors or to produce dimethylether (Arcoumanis et al. 2008). Methanol can, moreover, be reformed on board means of transport in a way that fits the use of $H_2$ in fuel cells (Ferreira-Aparicio et al. 2005).

Converting glycerol into methanol via syngas is now commercially applied. Furthermore, synthesis gas derived from glycerol may be turned – via the Fischer–Tropsch reaction – into hydrocarbons that may serve as diesel, petrol or kerosene (Scott et al. 2007; Simonetti et al. 2007; Valliyappan et al. 2008). Also, syngas may be used as a source of $H_2$ (Yazdani and Gonzalez 2007; Valliyappan et al. 2008). There are other options for converting glycerol into transport biofuels, too. Anaerobic fermentation may convert glycerol into ethanol and/or butanol (Coombs 2007; Yazdani and Gonzalez 2007). And glycerol may be converted into propanol, which can be mixed with conventional gasoline (Coombs 2007; Fernando et al. 2007).

### Biofuels from Complex Organic Feedstocks

Various processes generating transport biofuels start from complex organic feedstocks, including complete organisms or large chunks thereof, or from wastes such as sewage sludge, black liquor or household waste. The focus of such processes is often on lignocellulose. This is not surprising as the share of lignocellulose in all biomass has been estimated at about 50% (Claassen et al. 1999). Lignocellulose is a structural material of plants and a composite of lignin (a polymer composed of monolignols), cellulose (a glucose polymer) and hemicellulose (a polymer made up of diverse hexose and pentose sugars). The US Energy Law of 2008 stipulates that from 2016, transport bioethanol producers must switch to lignocellulosic feedstocks.

There are a wide variety of lignocellulosic feedstocks. Wood, wood waste, harvest residues, a variety of wastes and by-products originating in industries are relatively rich in lignocellulose (Prasad et al. 2007). It also has been suggested that natural grasslands can be exploited as a source for lignocellulosic feedstock (Tilman et al. 2006; Zhou et al. 2008). It is furthermore possible to grow lignocellulosic crops on plantations. Examples of species that are considered for this purpose are woody perennials, such as eucalyptus, poplar, willow and black locust, and grasses and other non-woody perennials, such as switchgrass, elephant grass, reed canary grass, *Miscanthus*, cardoon, reeds and Bermuda grass. Of these species, Bermuda grass and reed canary grass are currently used as forage for livestock (Boateng et al. 2007; Pahkala et al. 2008).

The relative amount of lignin in lignocellulose is source dependent. In nutshells, the percentage of lignin may be 30–40% and in rice straw about 5.5% (Prasad et al. 2007). Similarly, the composition of hemicellulose is source dependent. For instance, hemicellulose from agricultural residues or hardwood tends to be rich in pentose sugars, whereas such sugars are a minor component in hemicellulose from softwood (Hahn-Hägerdal et al. 2007).

Apart from lignocellulose, the lignocellulosic feedstocks also contain a variety of other compounds, both organic and inorganic in character. The latter, a fortiori, holds for complex wastes such as sludges from wastewater treatment plants or household wastes, which have been proposed as sources of transport biofuel (Ptasinski et al. 2002). There is also a relatively limited supply of cellulosic wastes that may be used for biofuel production, such as sludges from (virgin) paper production and paper recycling (Mabee and Roy 2003; Prasad et al. 2007; Marques et al. 2008).

There are several ways to generate substances that may serve as automotive biofuels from complex organic feedstocks. A first possibility, which applies both to biomass in general and to lignocellulose, is heating ('thermochemical treatment') to produce liquid biofuels. An option which has been exploited for centuries is the dry distillation or slow pyrolysis of wood. Apart from charcoal, methanol is an output (approximately 1–2% by weight) of slow pyrolysis of wood, which can in principle be used as a transport fuel (Reinharz 1985; Demirbaş 2001; Güllü and Demirbaş 2001; Huber et al. 2006). More recently, much attention has been given to fast and flash pyrolysis of biomass (Goyal et al. 2008). Fast and flash pyrolysis of biomass in principle produces charcoal or biochar, gas, organic fluids and water. The precise nature of the products and the relative shares of the different components can be varied, dependent on the character of the biomass, the presence of inorganic substances (especially metals), reactor design, temperature, heating rate, catalysts and reaction time (Bridgwater et al. 1999; Yang et al. 2004; Demiral and Şensöz 2006; Huber et al. 2006; Boateng et al. 2007; Dobele et al. 2007; Lange 2007; Müller-Hagedorn and Bockhorn 2007; Demirbaş 2008; Di Blasi 2008; Fahmi et al. 2008; Ros et al. 2009). The fluid produced by fast and flash pyrolysis contains water and a variety of organic compounds, the latter collectively called 'pyrolysis oil'. The pyrolysis oil tends to be unstable and to show polymerization reactions (Fahmi et al. 2008). It needs upgrading to serve as a basis for transportation fuel, for example, by hydrodeoxygenation, hydrogenation or treatment with zeolites (Huber et al. 2006; Esler 2007; Wang et al. 2008b). Such upgrading has proven difficult, and this has restricted the application of biomass pyrolysis technology (Wang et al. 2008b). It has also been proposed to view the pyrolysis oil as a basis for a biorefinery generating a number of chemicals besides transport fuel (Hayes 2008). As an alternative to fast and flash pyrolysis, a process has been proposed that combines pyrolysis with hydrogenation by a formic acid–alcohol mixture (Kleinert and Barth 2008). Also under development is deoxy-liquefaction (Goyal et al. 2008; Wang et al. 2008), converting lignocellulosic biomass into a liquid that tends to be richer in hydrocarbons than the liquids commonly produced by fast pyrolysis.

A second possibility to deal with complex organic feedstocks is based on gasification of biomass or lignocellulosic materials resulting in the formation of synthesis gas (containing relatively high percentages of CO and $H_2$). The formation of tar, and to a lesser extent char, and ash-related problems have emerged as problems in such gasification, necessitating major efforts in the field of optimizing gasification, tar reforming and syngas quality control and clean-up (Wang et al. 2008b). Using the water shift reaction, the amount of $H_2$ in synthesis gas may be maximized, and $H_2$ and the other main product of the water shift reaction ($CO_2$) can be separated by processes such as pressure swing adsorption, membrane separation and cryogenic separation (Ferreira-Aparicio et al. 2005; Andersson and Harvey 2006; Haryanto et al. 2007; Barelli et al. 2008; Florin and Harris 2008; Wang et al. 2008b). It is also possible to subject syngas (after clean-up) to catalytic methanation, generating synthetic natural gas (Felder and Dones 2007).

Alternatively, conversion is possible into liquids (biomass-to-liquids or BTL biofuels). One option is the use of synthesis gas to produce oxygenates such as methanol (Reed and Lerner 1973; Demirbaş 2001; Ptasinski et al. 2002) and dimethylether (Joelsson and Gustavsson 2008). Producing ethanol from syngas is also possible but is as yet not very efficient (Subramani and Gangwal 2008). Still another option is to use the Fischer–Tropsch reaction, after enrichment of syngas with hydrogen, to generate hydrocarbons (Dietenberger and Anderson 2007), or the methanol-to-synfuel synthesis to produce hydrocarbons (Takeshita and Yamaji 2008). The latter can be conveniently applied in diesel or Otto motors (Reinhardt et al. 2006) or in airplanes (Esler 2007). There are also bacteria that can convert synthesis gas into ethanol, and these are currently researched for use in biofuel production (Henstra et al. 2007; Tollefson 2008). Low conversion rates, product inhibition and problems in maintaining optimum conditions have for a substantial time prevented commercialization of this approach (Wang et al. 2008b), but such problems have now apparently been solved to the extent that a pilot plant has been announced (Ashley 2008).

Thirdly, cellulose and hemicellulose present in lignocellulose may be enzymatically converted into ethanol or butanol, to be applied in, for example, Otto motors (Sánchez and Cardona 2008; Qureshi et al. 2008a, b). This requires separating hemicellulose from lignin, hydrolysis of cellulose and hemicellulose into sugars and fermentation of the sugars generated by hydrolysis (Lynd 1996; Lachke 2002; Palmarola-Adrados et al. 2005; Gray et al. 2006; Angenent 2007; Prasad et al. 2007; Gomez et al. 2008; Sánchez and Cardona 2008; Qureshi et al. 2008a, b).

Hydrolysis of cellulose generates glucose, which can be converted into ethanol. Important among the hydrolytic products of hemicellulose is often xylose, a 5-carbon sugar (Fortman et al. 2008). Xylose can be converted into ethanol by fermentation as follows:

$$3\,\text{D-xylose}\,(C_5H_{10}O_5) \rightarrow 5\,\text{ethanol} + 5\,CO_2\,.$$

Micro-organisms such as *Pichia stipitis* and genetically modified *Escherichia coli* are able to perform the fermentation of xylose (Rubin 2008). Minor sugars originating in cellulose and hemicellulose are arabinose, rhamnose, glucose, galactose and mannose, which can be converted into ethanol, too (Numan and Bhosle 2006; Fortman et al. 2008; Hayes 2008). It is also possible to ferment $C_6$ and $C_5$ sugars into a mixture of acetone, butanol and ethanol (Jones and Woods 1986; Qureshi et al. 2008c). Process design tends to be focused on a limited number of lignocellulosic feedstocks for which the process is optimized (Olofsson et al. 2008). In practice, the separation of hemicellulose from lignin currently causes most problems, which are in part linked to the heterogeneous structure of lignin polymers (Gomez et al. 2008; Wackett 2008). Building cell walls involves many enzymes (McCann and Carpita 2008), and it may well be that a combination of enzymes may be necessary for their deconstruction in a way that is optimal for the next step of biofuel production: saccharification. However, most processes currently studied for near-term application rely on the use of rather brute physico-chemical force to separate the constituents of lignocellulose (which negatively impacts overall energy efficiency and the environmental burden). Examples are: the use of acid (whether or not combined with ionic liquid), steam explosion (sometimes combined with oxidation), high-pressure hot water treatment, treatment with alkaline peroxides and ammonia fibre explosion (Huber et al. 2006; Gomez et al. 2008; Li et al. 2008; Sørensen et al. 2008; Qureshi et al. 2008c). The difficulty of separation varies for different plant species (Buranov and Mazza 2008). Coniferyl lignin appears, for instance, more recalcitrant so far against physico-chemical methods of separation than syringyl lignin (Anderson and Akin 2008). And the presence of oxidatively coupled esterified or etherified ferulic acid residues has also been reported to inhibit separation (McCann and Carpita 2008).

Proposals to overcome the hurdles to separation of lignin and cellulose and hemicellulose include: application of lignin-degrading white rot fungi of microorganisms derived from termite guts, of *Clostridium phytofermentans*, and pre-treatment with phenolic esterases (Warnick et al. 2002; Anderson and Akin 2008; Rotman 2008; Weng et al. 2008). Also, it has been suggested to use lignases and to convert degraded lignin into transport biofuels (Blanch et al. 2008). Furthermore, there have been proposals to downregulate lignin biosynthesis in plants by genetic modification to ease the release of cellulose and hemicellulose and ultimately sugars from plants (Chapple et al. 2007; Wackett 2008). Such downregulation has led to plant characteristics that are unsuitable for biofuel crops, such as increased susceptibility to fungi, dwarfing and the collapse of vessels in xylem (Weng et al. 2008). Dwarfing has been linked to the simultaneous inhibition of flavonoid production (McCann and Carpita 2008). There have been new proposals for genetic modification, focusing on changes in lignin polymer structure and monolignol polymerization (Weng et al. 2008), but it is as yet not clear whether this approach will lead to suitable biofuel feedstocks.

Degradation of hemicellulose may also be difficult. Hemicelluloses appear so far refractory against saccharification when esterified by ferulic or coumaric acids (Anderson and Akin 2008). And enzymatic hydrolysis of cellulose to fermentable sugars currently requires, per kg ethanol produced, 40–100 times more enzyme than the hydrolysis of starch (Eijsink et al. 2008). This has led to the proposal to include glycosyl hydrolases into plants by genetic modification (Taylor et al. 2008). Hydrolysis and fermentation of cellulose and hemicellulose can be done in a two-step process (e.g. Zanichelli et al. 2007; Hayes 2008), with one hydrolytic and one fermentative step. When dilute acid is used for the pre-treatment of lignocellulosic biomass, there is often much hydrolysis of cellulose and hemicellulose. At higher temperatures, dilute acid treatment may also lead to much hydrolysis of cellulose (Hayes 2008). A variant of treatment with dilute acid may be used to generate substantial amounts of the platform chemicals furfural and levulinic acid, in line with biorefinery concepts (Hayes 2008). Alternatively, hydrolytic enzymes produced by micro-organisms may be used (Lynd et al. 2002; Demain et al. 2005; Desvaux 2005).

The two steps in the conversion of cellulose and hemicellulose to ethanol may also be combined in a one-step process: simultaneous saccharification and fermentation (SSF). Simultaneous saccharification (hydrolysis of cellulose and hemicellulose giving rise to sugars) and fermentation by micro-organisms is often preferred as it is associated with shorter residence times and potentially higher yields and lower costs (Ballesteros et al. 2004; Demain et al. 2005; Huber et al. 2006; Angenent 2007; Marques et al. 2008). In simultaneous saccharification and fermentation, saccharification is the rate-limiting step. Inhibition of fermentation by substances formed during pre-treatment and hydrolysis is a problem. Inhibitory compounds formed during pre-treatment and hydrolysis include salts, phenols, furfural, cinnamaldehyde, *p*-hydroxybenzaldehyde, lignin monomers and syringaldehyde (Zanichelli et al. 2007; Qureshi et al. 2008c; Sánchez and Cardona 2008; Royal Society 2008). The presence of inhibitors often necessitates the 'detoxification' by physical, chemical or biological methods. Another option is the use of fermenting organisms that are more tolerant to inhibitors (Hayes 2008; Olofsson et al. 2008).

Processes converting cellulose and hemicellulose into ethanol have as yet relatively low sugar-to-ethanol efficiencies, if compared with the well-established starch- or sucrose-to-ethanol conversion processes (Chang 2007; Hahn-Hägerdal et al. 2007; Olofsson et al. 2008). In view of the problems in converting lignocellulosic feedstocks into alcohol, there is a lively search for improvements, if not a 'technological breakthrough' or a 'superbug' that is able to perform the task of converting lignocellulose into ethanol with sufficient efficiency (Eijsink et al. 2008; Gomez et al. 2008; Rotman 2008).

In North America and Scandinavia, ethanol from woody lignocellulose has been, and still is, produced as a by-product of sulphite pulping for the paper industry (McElroy 2007). Wood hydrolysate is in this case converted into ethanol by yeast-based fermentation. Low nutrient concentrations, a large proportion of xylose and the presence of fermentation inhibitors have limited the efficiency thereof, and there are proposals for the optimization of sulphite liquor fermentation (Helle et al. 2008).

In Russia, there is a long-standing, large-scale, yeast-based production of ethanol from sugars obtained from wood chips hydrolyzed at elevated temperature by treatment with concentrated sulphuric acid (Bungay 2004; Zverlov et al. 2006). In Brazil, Europe and the USA, there are pilot plants producing ethanol from lignocellulose or components thereof such as cellulose and hemicellulose (Wheals et al. 1999; Bryner 2007a). Large-scale plants are under construction and consideration. In part, ethanol production from lignocellulose in such plants is combined with ethanol production based on sugar or starch.

Alternatively, a bacterial fermentation process for the production of the biofuel butanol from lignocellulose may be considered (Zverlov et al. 2006; Ezeji et al. 2007; Qureshi et al. 2008a, b). This process was used during the twentieth century in the Soviet Union for the fermentation of hydrolyzed lignocellulosic wastes (Zverlov et al. 2006). In this case, lignocellulose was hydrolyzed by treatment with high concentrations of sulphuric acid, and the hydrolysate was fermented in combination with the fermentation of starch. $H_2$ originated as a by-product (Zverlov et al. 2006). Also, processes producing butanol from lignocellulose based on treatment with dilute acid followed by enzymatic treatment and fermentation (simultaneous saccharification and fermentation) of harvest residues have been proposed (Ezeji et al. 2007; Qureshi et al. 2008a, b), as have been processes to convert complex organic feedstocks into mixtures of alcohols using mixtures of fermentative bacteria (Bagajewicz et al. 2007). Finally, there is research into the possibility of enzymatically converting lignocellulose into fatty acid ethyl esters (Royal Society 2008).

Fourthly, methane may be produced from complex organic materials. Methane is also the molecule that makes natural gas a fuel, and natural gas supplies currently about 3% of primary energy for transport (de la Rue du Can and Price 2008). In 2004, there were about 3 million motorcars powered by natural gas, usually bivalent vehicles able to drive on compressed natural gas and gasoline (Dondero and Goldemberg 2005; Janssen et al. 2006). Substantial use of vehicles powered by natural gas is found in Argentina (world leader with about 800,000 of such vehicles by 2005), India, Pakistan, Brazil, the USA and some countries in the European Union, such as Italy (Janssen et al. 2006). Large-scale application of methane in cars is dependent on a good refuelling infrastructure (Janssen et al. 2006). Natural gas is also used in ship and on-farm transport (Royal Society 2008; Börjesson and Mattiasson 2008). Alternatively, methane may be converted into liquid fuels using the Fischer–Tropsch reaction or via a process with ethylene as an intermediary (Hall 2005). Currently, use of methane from natural gas in the Fischer–Tropsch synthesis of hydrocarbons is applied in diesel production, and this application is expected to increase in the future (Bagajewicz et al. 2007; Bryner 2007b; Takeshita and Yamaji 2008). Methane can furthermore be converted into methanol (Huber et al. 2006; Cantrell et al. 2008).

The use of methane in transport and the production of other transport fuels may be extended to biogenic methane (Murphy and McCarthy 2005; Börjesson and Mattiasson 2008; Lehtomäki et al. 2008). Above, the production of synthetic natural gas from syngas has already been referred to. Methane can also be produced from a wide variety of biomass and biomass-derived materials, including complex wastes, using mixed cultures of micro-organisms in anaerobic reactors (Murphy and McCarthy 2005; Kleerebezem and van Loosdrecht 2007; Bocher et al. 2008; Ros et al. 2009). It has been proposed to use marginal lands for the large-scale growth of feedstocks and convert those into methane in decentralized biogas reactors (Schröder et al. 2008). A variant of this approach has been suggested that also allows for the bioconversion of $CO_2$ to $CH_4$ (Alimahmoodi and Mulligan 2008). Landfills can also be exploited for methane production. Before application in transport, $CH_4$ production from biomass should be followed by upgrading. The extent of upgrading necessary varies, depending on the methane source. More upgrading is usually needed for methane from refuse in landfills and sewage sludge than for methane from manure (Rasi et al. 2007). Upgrading partly serves to remove compounds that may negatively affect engine performance or emissions, such as halogenated compounds, siloxanes, $H_2S$ and $NH_3$ (Ferreira-Aparicio et al. 2005; Rasi et al. 2007). Upgrading may also aim to increase methane content.

#### Hydrocarbons ('Biocrude') from Terrestrial Plants

In the 1930s, there were some efforts to cultivate *Euphorbia*, producing hydrocarbons for biofuel production (Kalita 2008). Subsequently, there has been substantial research regarding plants producing latex which may be cracked to yield transport biofuels. The *Euphorbia lathyris* did relatively well in this respect and has been calculated to yield about 48 MJ $ha^{-1}$ $year^{-1}$ in biofuel: 26 MJ as hydrocarbons and 22 MJ as ethanol (Kalita 2008). As will be shown in Chap. 2, such energetic yields are relatively low compared with biofuels from current terrestrial crops, and there is no current commercial application of 'biocrude' from terrestrial plants.

### 1.5.2 Biofuels from Aquatic Biomass

A variety of algae are currently cultivated commercially, especially for applications in food and feed production, but also for other applications such as fertilizer and the production of materials. Also, there is limited harvesting of uncultivated algae (Critchley et al. 2006). There have been proposals to exploit aquatic biomass for the production of biofuels.

### Marine Phytobiomass

Most of the surface of the Earth consists of seas, mainly oceans. A variety of proposals exist to exploit the seas for the production of biofuels. Macroalgae, macrophytes and microalgae have been considered in this context. Microalgae include both prokaryote and eukaryote photosynthetic micro-organisms. In the context of exploiting macrophytes, floating man-made structures to cultivate the *Macrocystis pyrifera* (giant kelp) have been proposed (Wilcox 1982; Bungay 2004). Varieties of the brown seaweed *Laminaria*, which is currently harvested for food (Chopin et al. 2001), have been suggested as a convenient source of carbohydrates to be converted into ethanol (Horn et al. 2000). The highly salt-tolerant microalga *Dunaliella*, which, for instance, occurs in the Dead Sea, has also been proposed as a source of transport biofuel (Ben-Amotz et al. 1982).

However, there is a major snag regarding the proposal to use the sea for the production of algae, which may serve the supply of transport biofuels. Actual phytobiomass in the seas is in the order of 1–2% of total global plant carbon. Photosynthesis in the seas is much higher (in the order of 40–50% of total photosynthesis; Rosing et al. 2006), but most of the photosynthetic yield (approximately 80–88%) is quickly consumed. In the case of microalgae, consumption is mainly by zooplankton ('grazers'), while 2–10% is subject to viral lysis (Wilhelm and Suttle 1999). Thus, substantial direct appropriation of the products of photosynthesis by humans in the seas in general would necessitate a major overhaul of the marine food web. For the successful growth of desired microalgae, probably dramatic changes in ocean composition, such as a switch to much higher salinity, may be required (Sawayama et al. 1999; Joint et al. 2002; Ugwu et al. 2008). Large-scale exploitation of macroalgae is cumbersome. Proposals to exploit the giant kelp *Macrocystis* require pumping of deep seawater to the ocean surface, massive man-made structures to support kelp growth and regular replanting (Bungay 2004).

### Near-Shore Marine Phytobiomass

Near-shore perspectives for exploiting macroalgae may be different. Firstly, there are cases where macroalgae have developed into a pest because of eutrophication (Morand and Merceron 2005). In some of these cases, significant amounts of these macroalgae are currently collected and landfilled. For instance, in Europe, this happens in parts of the Venice Lagoon, the Ortbetello Lagoon, the Bay of Brittany and the Peel Inlet, with collected amounts in the order of $10^3$–$10^4$ Mg per year (Morand and Merceron 2005; Bastianoni et al. 2008). Such macroalgae may be used for biofuel production. However, when nutrient emissions are reduced, macroalgal primary production will also be diminished. Secondly, there is already major near-shore cultivation of macroalgae mainly for food and feed (Wikfors and Ohno 2001; Critchley et al. 2006). Eutrophication of coastal waters is conducive to yields, and there is also intentional addition of nutrients to further production (Neushul and Wang 2000). Also, it has been suggested to combine cultivation of macroalgae with nutrient emissions from marine animal aquaculture (Wikfors and Ohno 2001; Chopin et al. 2001; Troell et al. 2006).

### Microalgae from Open Ponds and Bioreactors

While aiming at transport biofuels, the growth of microalgae with high levels of oil (triacylglycerol) followed by lipid extraction has drawn most attention (Scragg et al. 2002; Wijffels 2008; www.oilgae.com; Dismukes et al. 2008; Liu et al. 2008). Such lipids can then subsequently be converted into replacements for fossil fuels, in ways similar to vegetable oils from terrestrial plants. There is currently some use of biodiesel based on algal oils, as pointed out above. There have also been proposals to convert algal biomass into methanol via synthesis gas or into bio-oil via pyrolysis (Hirano et al. 1998; Sawayama et al. 1999). Strains of the photosynthetic microalga *Botryococcus braunii* may contain and secrete substantial amounts of isoprenoid hydrocarbons: *n*-alkadienes and trienes, methylated squalenes and terpenoids (Guschina and Harwood 2006). When subjected to catalytic cracking, these hydrocarbons can be converted into transport biofuels (Banerjee et al. 2002). It has also been suggested that an intermediate in the synthesis of isoprenoids by *Botryococcus braunii* (isoprenylpyrophosphate) may be converted into isopentanol, which may be used as a gasoline additive (Fortman et al. 2008). The slow growth of *Botryococcus braunii* has not been conducive to its application.

Microalgae may be produced in open ponds converting solar irradiation into biomass which may be harvested and converted into biofuels. Open ponds used for growing microalgae are man-made structures (made from, for example, plastic or concrete) with 10–20 cm of water that are subjected to circulation and mixing (Chisti 2007). Closed systems ('bioreactors') have also been proposed for the purpose of growing photosynthetic micro-organisms to produce transport biofuels (Chisti 2007, Wijffels 2008). In closed systems, heterotrophs, organisms that graze on algae (zooplankton) and viruses can be excluded, and monocultures of desirable species can be maintained. In open ponds, sustained generation of a specific photosynthetic micro-organism with relatively little contamination of other species and subject to low heterotrophic conversion would seem only possible under extreme circumstances, such as very high salinity and/or a high pH (Joint et al. 2002; Ugwu et al. 2008). Sustained open pond production has been successful for a limited number of algae such as *Spirulina*, *Chlorella* and *Dunaliella* (grown at high pH and/or NaCl concentrations). For other organisms, most growth can take place in a closed bioreactor, which then may be eventually followed by a short period in an open pond (Huntley and Redalje 2007).

**Freshwater Macrophytes**

In fresh waters, there has been an emergence of invasive macrophytes with high primary production per hectare. Increased levels of nutrients ('eutrophication') and the import of macrophytes from other continents have been conducive to this emergence (Gassmann et al. 2006; Gunnarsson and Petersen 2007). Among these macrophytes, the water hyacinth (*Eichhornia crassipes*) has been studied in the context of biofuel production (Gunnarsson and Petersen 2007; Malik 2007). The water hyacinth has emerged as a major invasive organism ('pest') in tropical freshwater systems especially outside its natural range (South America). Water hyacinth biomass forms floating mats which interfere with shipping, power generation, drinking water production and irrigation, are detrimental to fish stocks and may be conducive to a number of infectious diseases (Odada and Olago 2006; Gunnarsson and Petersen 2007; Malik 2007). Due to these negative impacts, there are efforts to reduce the presence of *Eichhornia crassipes* in tropical surface waters, which have met with at least some success (Odada and Olago 2006). The need to control the water hyacinth is evidently at variance with high yields, but when the water hyacinth generates substantial amounts of floating biomass, energetic use thereof may be considered (Gunnarsson and Petersen 2007; Malik 2007).

## *1.5.3 Hydrogen Production Mediated by Micro-Organisms*

The use of a variety of photosynthetic organisms has been proposed that directly or indirectly biocatalyse the splitting of water into $H_2$ and $O_2$ (Melis and Happe 2001; Hallenbeck and Benemann 2002; Nath and Das 2004; Hahn et al. 2007; Hankamer et al. 2007). The production of hydrogen from wastewaters or carbohydrates by $H_2$-producing bacteria has also been proposed (Van Ginkel et al. 2005; Rupprecht et al. 2006; Wongtanet et al. 2007; Jones 2008). The latter approach to generating $H_2$ so far has a poor conversion efficiency (Jones 2008).

Melis and Melnicki (2006) have suggested the combined production of hydrogen by $H_2$-producing bacteria and photosynthetic algae, and Westermann et al. (2007) have proposed biorefineries producing ethanol and hydrogen. Furthermore, there have been proposals to electrolytically generate hydrogen from wastes in the presence of microbes (Stams et al. 2006; Dumas et al. 2008).

The methods to produce $H_2$ with the help of micro-organisms would often require closed systems (Rupprecht et al. 2006). This is necessary for the capture and removal of $H_2$, which may inhibit $H_2$ production, maintain anaerobic conditions and be highly conducive to limiting infection by unwanted organisms, but limits the scope for large-scale production.

## 1.6 Biofuel-Based Electricity for Transport

As pointed out before, about 1% of primary energy use for transport worldwide concerns electricity (de la Rue du Can and Price 2008). Moreover, especially because the electromotor is more efficient than internal combustion engines, electric traction with electricity derived from power stations is relatively fuel efficient (e.g. Reijnders and Huijbregts 2007).

In Sect. 1.3, the types of organic materials that are currently used in electricity production have been outlined. It has also been suggested to use herbaceous crops that generate dried down biomass, such as horseweed and sunflower (Kamm 2004) for electricity production. Substantial expansion of biofuel-based electricity production for transport is dependent on the social acceptability of electric traction in cars. Interestingly, there have been periods in which the social acceptability of electric traction has been high for types of car transport now dominated by internal combustion engines. In 1899/1900, electric motorcars outsold other types of cars in the USA (Høyer 2008), electric taxis were then highly popular and between 1900 and 1920, electric vans were important in intra-urban and suburban transport of a variety of goods in the USA (Mom 1997; Mom and Kirsch 2001). All in all, the 1880–1925 period was a golden age for electric cars in the USA and parts of Western Europe (Mom 1997; Høyer 2008). In the 1940s, vans used for the German postal services and for milk and bread delivery in Britain were usually electric (Høyer 2008). Still, electric cars meet a substantial demand of large fleet owners in urban settings (e.g. as post office and street cleaning vans). Cars powered by electricity from power stations have, however, only had very limited success among individual users (Gjoen and Hard 2002). Opinions diverge about their future potential. Some view the re-emerging interest in electric cars as an episode in a series of unsuccessful attempts to substantially increase the use of electric cars. Others predict that there will be a rapid and fast increase in the use of electric cars. Lache et al. (2008) suggest a rapidly increasing market share for plug-in electric vehicles in Europe, with lithium batteries as key enabling technology. Others suggest that large socio-cultural changes, major technical changes and substantial financial incentives are necessary to make plug-in, battery-powered, all-electric traction for cars much more popular in the future (Delucchi and Lipman 2001; Gjoen and Hard 2002; Chalk and Miller 2006; Høyer 2008).

When electric traction gets a much larger share in road transport, it is likely that two technologies will contribute significantly to its success. The first is better batteries. Prime candidates are currently lithium ion batteries, which for a specified electrical performance are, over their life cycle, less of an environmental burden than competing batteries, such as the lead–acid and nickel-based batteries (Matheys et al. 2007). The second is plug-in hybrid cars, which in their life cycle energy use may have an advantage over current hybrid cars (Samaras and Meisterling 2008).

## 1.7 Recent Development of Transport Biofuel Production: Volume, Costs and Prices

### *1.7.1 Volume of Biofuel Production*

Some companies operating means of public transport which use electric traction have opted for 'green electricity', which may include biomass-based electricity production. No data have been found that allow a worldwide estimate of biomass-based electricity in electric traction. Still, the production of bioethanol apparently accounts for most of the current volume of transport biofuel production. The focus in the USA is largely on ethanol made from cornstarch, and in Brazil, it is mostly on ethanol made from sugar cane. China has also emerged as a major producer of bioethanol, preferentially from sugar cane, cassava and yams (Cascone 2007), and so has the European Union, producing bioethanol from wheat and sugar beet (Berndes and Hansson 2007). India, Russia, Southern Africa, Thailand and the Caribbean are emerging as important producers of ethanol as a transport fuel (Cascone 2007; Szklo et al. 2007; Barrett 2007; Amigun et al. 2008; Nguyen et al. 2008). Estimated bioethanol production volumes for 2006 in the main production areas are given in Fig. 1.2 and sum up to a world estimate of $51 \times 10^6$ m$^3$. The estimated world production of bioethanol in 2007 was $54 \times 10^6$ Mg (Monfort 2008).

The worldwide production of biodiesel in 2006 was probably in the order of $6.4 \times 10^6$ Mg, with the share of the European Union being approximately 77% and of

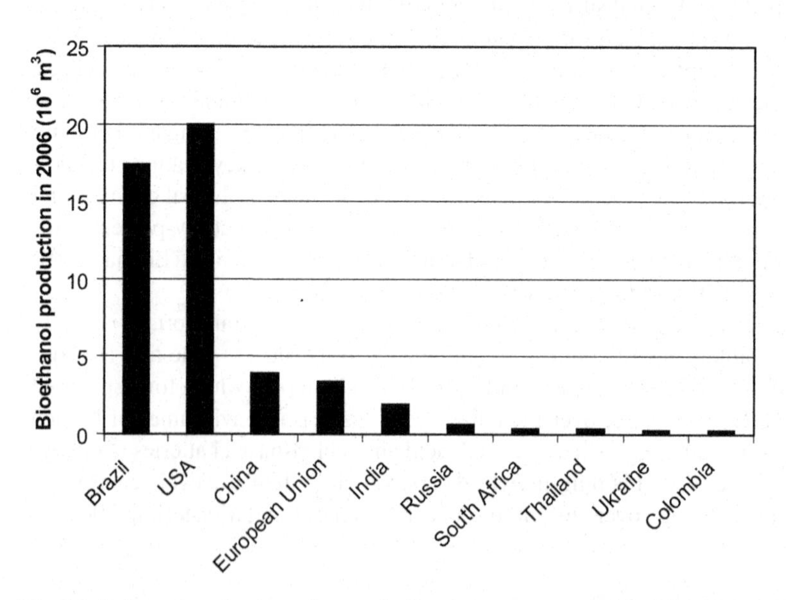


**Fig. 1.2** Estimated production volumes of ethanol as a transport fuel in 2006 (Licht 2006; Antoni et al. 2007; Szklo 2007; Sanchez and Cardona 2008)

the USA approximately 12% (Canakci and Sanli 2008). In 2007, estimated biodiesel production was approximately 7.6–8 $\times 10^6$ Mg (Von Braun 2008; Monfort 2008). Recently, there has been rapid growth especially in Argentine biodiesel production capacity. Argentine biodiesel production is expected to grow to about $3.5 \times 10^6$ Mg per year by 2008/2009 (Lamers et al. 2008). Biodiesel production on the basis of castor oil is expanding in Brazil, though it has been argued that in view of properties and price, castor oil is unlikely to be competitive with palm oil and rapeseed oil (Mathews 2008b; Scholz and da Silva 2008). India, Malaysia, Nicaragua, and several Pacific island states are also involved in substantial biodiesel production (Grimm 1999; China Chemical Reporter 2007; Cascone 2007; Cloin 2007; Fairless 2007; da Costa et al. 2007; Runge and Senauer 2007). Much of the biofuels produced are for domestic use, but increasingly, biofuels are traded internationally. Brazil and Argentina are, for instance, emerging as major transport biofuel exporters.

Most of the biofuels which were produced for transport applications in 2006 were based on substances that are also applied as foodstuffs. Such biofuels have been called 'first generation biofuels'. Biofuels can also be produced on the basis of other substances, such as lignocellulosic feedstocks and oils that are not foodstuffs. This category of biofuels has been called 'second generation biofuels'. However, ethanol production as a by-product of the sulphite pulping process, Russian lignocellulose-based ethanol production and the application of *Jatropha* oil for biodiesel production have evolved contemporaneous to or even before first generation biofuels such as cornstarch-based bioethanol and rapeseed biodiesel (Grimm 1999; Zverlov et al. 2006; McElroy 2007). Also, algal biofuels are often called second generation biofuels, but several of the algae considered for this purpose have current applications as food.

Moreover, as it is apparently felt that second generation is somehow better than first generation, the former designation is used in strange ways, for instance for deoxygenated and hydrogenated edible vegetable oils (Rantanen et al. 2005; Mikkonen 2008). For these reasons, the designations first and second generation will not be further used. It seems likely that biofuels made from substances that may also serve as food or feed will dominate the supply in the near future. Plans for other types of biofuel, when implemented, will by 2010 probably not be able to supply more than 1% of overall biofuel production, and such biofuels are unlikely to allow for large-scale replacement of biofuels from substances such as sugar, starch and edible vegetable oil before 2020 (Gibbs et al. 2008; OECD 2008).

Growth of biofuel production and/or consumption is foreseen in a number of countries. In Brazil, sugar cane production, as feedstock for ethanol, is expected to grow from $425 \times 10^3$ Mt in 2006 to $728 \times 10^3$ Mt in 2012 (Macedo et al. 2008), while the mandate for Brazilian biodiesel is set at 5% for 2013. In 2007, the USA mandated a growth of bioethanol production from 4.7 billion gallons in 2007 to 36 billion gallons by 2022. In 2008, there were, however, calls for revision of this target in the US Congress (Doering 2008). In 2008, Canada mandated a 5% ethanol blend in gasoline by 2010 and a 2% biodiesel blend in on-road diesel by 2012. The European Union in 2007 suggested a 10% share of biofuels in transport fuels by 2020, which in 2008 was hotly debated.

### *1.7.2 Costs of Biofuel Production and Biofuel Prices*

The cost of biofuels is a longstanding topic of discussion, especially in relation to the costs of competing fossil fuels. Two types of costs are involved: the costs of producing biofuels and costs that users have in adapting to biofuels. The latter costs are highly variable. Co-firing wood pellets in power plants to power electric trains has low adaptation costs, and the same holds for adding low percentages of biofuel to conventional gasoline and diesel. However, for instance, switching from diesel and gasoline to (biofuelled) electric traction is a major operation. Here we will further focus on the costs of transport biofuel production.

For producers, there again are two types of costs. Firstly, there are costs borne by the producer. Secondly, there are external costs or externalities (Pigou 1920). External costs are (fuel-linked) costs that are not reflected in actual prices. Such costs are associated with negative environmental impacts, including negative impacts of air pollution on health (Johansson 1999) and on ecosystems, and the future availability of natural resources. But there are also other external costs associated with fuels, such as the costs of strategic stockpiling and in the case of mineral oil, military costs involved in safeguarding the supply (Zaldivar et al. 2001; Delucchi and Murphy 2008). Such costs are substantial and may vary strongly between fuels (Johansson 1999). However, as long as governments do not succeed in fully 'internalizing' such external costs, they will have very little impact on economic decision making. So here, only costs borne by the producer will be considered. Figure 1.3

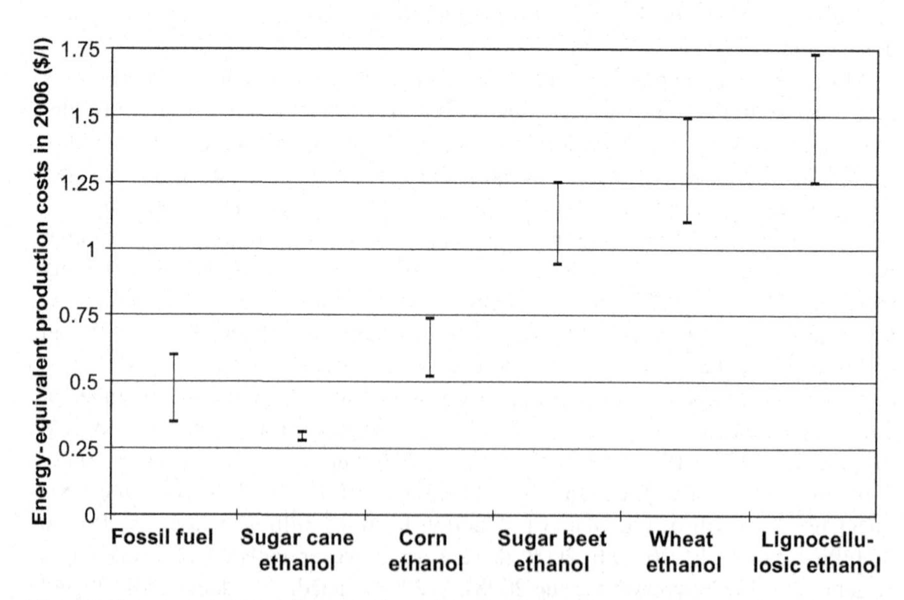


**Fig. 1.3** Fuel costs in US dollars per litre of fossil-fuel-based transport fuel and energetically equivalent amounts for bioethanol varieties in 2006, recalculated from data in Licht 2006; Szklo et al. 2007 and Royal Society 2008

shows cost estimates for per-litre, fossil-fuel-based transport fuels and the energetic equivalent thereof for varieties of bioethanol in 2006.

What have also emerged are major regional differences in biofuel production costs, probably linked to differences in costs of land and labour and yields of feedstocks. This is shown by Fig. 1.4, which gives biodiesel production costs for 2006 while not taking account of external costs. All prices in Fig. 1.4 refer to biofuels from terrestrial plants. Estimates about the costs of large-scale production of biodiesel from algal oil are in the order of US $2.90 per litre (Chisti 2007), whereas transport biofuels from cultivated macroalgae would even be more expensive as the price range of the latter is more in line with their use as a delicacy (Neushul and Badash 1998; Buschmann et al. 2001). The cost of biodiesel made from used cooking oil and animal fats has been estimated at about US $0.22–0.74 per litre (Johnston and Holloway 2007; Canakci and Sanli 2008; Royal Society 2008).

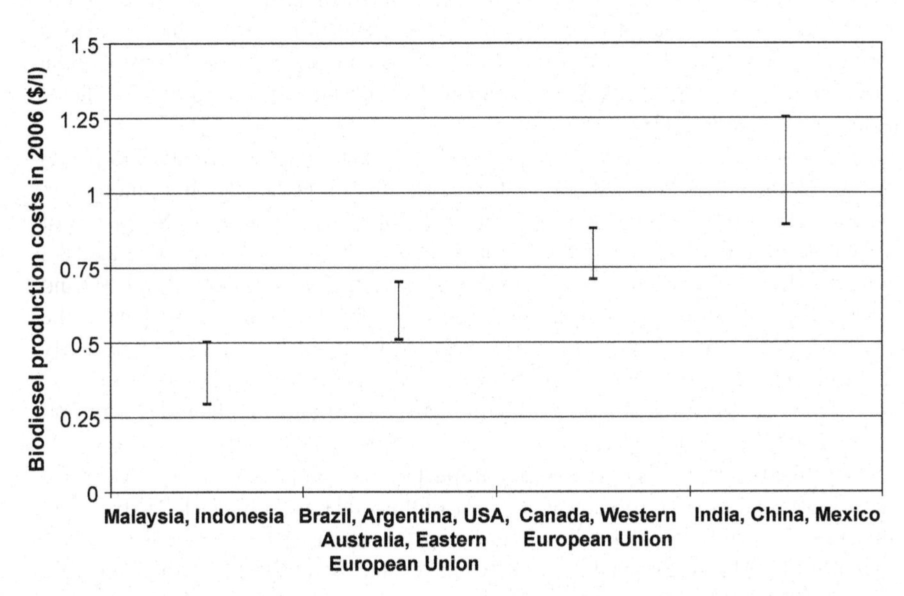


**Fig. 1.4** Production costs in US dollars per litre of biodiesel in early 2006 (Demirbaş 2007; Johnston and Holloway 2007)

In Brazil, as indicated by Fig. 1.3, during 2006, ethanol from sugar cane could compete with fossil-fuel-based transport fuels, but in the USA and the European Union, ethanol prices in 2006 were such that they were not competitive with gasoline when external costs of fuels and fuel production were not included. By mid-May 2008, the situation was changed. Then, when costs were compared, corn-based ethanol in the United States was competitive with fossil gasoline (Westhoff 2008).

Figure 1.4 shows that in 2006, biodiesel from vegetable oil produced in Europe or the USA was not competitive with fossil-fuel-based transport fuels, but that in Malaysia and Indonesia, it was competitive when external costs were not included. Costs for different types of biofuel partly reflect differences in maturity of the production process. The relatively low price for sugar-cane-based ethanol partly reflects a long learning curve (Goldemberg et al. 2004), though there is still scope for additional cost reduction (Hayes 2008; Macedo et al. 2008). Ethanol from lignocellulose is in an early stage of development, and there is much scope for cost reduction (Hayes 2008). It has been claimed that lignocellulosic ethanol can ultimately become competitive with ethanol from corn (Frederick et al. 2008; Lynd et al. 2008).

Both mineral oil and biomass prices are subject to change, and this may strongly affect the relative attractiveness of biofuels. For instance, Brazilian ethanol production did well when oil prices were relatively high, but demand slumped when such prices were low. It is often argued that mineral oil prices will in the future probably remain relatively high, which seems to bode well for the competitive position of biofuels. However, experience shows that predictions as to when biofuels become competitive with fossil fuels are subject to a major uncertainty – the prices of feedstocks. This is exemplified by the situation in 2008. Crude oil prices temporarily achieved price levels in the order of greater than US $100 per barrel, but biofuel feedstock prices also rose sharply. So, for instance, in 2008 and dependent on feedstock, the biodiesel unit price was 1.5–3 times higher than that of mineral-oil-derived diesel (Canakci and Sanli 2008).

For feedstocks that may also serve as a basis for food, major changes in prices are well known from the past. For instance, coconut oil prices varied by more than a factor of seven over the last 40 years (Cloin 2007). The nominal (US $) price of vegetable oil changed by about a factor of two in the 1997–2000 period, and the nominal (US $) price of wheat increased by about a factor of two between 1999 and 2006 (OECD-FAO 2007). And over the February 2007 to February 2008 period, the price of palm oil roughly doubled (www.palmoil.com). From early 2006 to early 2008, the price of US corn went from US $87 per metric ton to US $217 per metric ton (Tyner 2008). Price volatility may increase due to climate change (Eaves and Eaves 2007; Lobell et al. 2008).

High feedstock prices have a strong impact on biofuel prices. In the mid-1990s, the cost of biodiesel feedstock was 60–75% of the total cost of biofuel, and by 2008, this was 85%, with a $0.20 per litre biodiesel price increase when the feedstock price increased by US $0.22 per kilogram (Canakci and Sanli 2008). Similar changes occurred for starch- and sugar-based alcohols (Claassen et al. 1999; Qureshi and Blaschek 2001; Huber et al. 2006; Demirbaş 2007; Koizumi and Ohga 2007; You et al. 2008). Furthermore, changes in prices of by-products do not necessarily favour the profitability of biofuel projects. Mainly due to expanding biodiesel production, a glycerol glut has emerged, which has negatively affected glycerol prices (Willke and Vorlop 2004; Yazdani and Gonzalez 2007). In 2007, glycerol prices were lowered to a level well below that previously used in the calculation of biofuel prices (e.g. Francis et al. 2005; Huber et al. 2006). That co-products of biofuel production may be subject to change may have consequences for prospective biofuel prices. For instance, the relatively low price for microalgal biodiesel suggested by Huntley and Redalje (2007) is dependent on the current high value for the co-product astaxanthin, but the price of astaxanthin may plummet if the production of algal biodiesel were to expand greatly (Vasudevan and Briggs 2008).

Further rapid expansion of biofuel production has also been argued to contribute to relatively high prices for the major commodities from which biofuels are made: crops for vegetable oil, starch and sugar (Runge and Senauer 2007; Daschle et al. 2007; Naylor et al. 2007). Actual predictions about future prices for vegetable oil, starch and sugar crops are extremely variable (Naylor et al. 2007). So firm predictions as to the relative future costs of fossil and biofuels, if based on the crops from which they are currently largely made, are hard to make. However, when biofuel production from food crops becomes large scale, prices of crops that serve as major biofuel feedstocks are expected to follow the price of crude oil, when corrected for the energy content of the biofuel (Naylor et al. 2007; Westhoff 2008).

It has been argued that the situation will be different when lignocellulose is used as a basis for transport biofuel production. Here, estimates of feedstock costs are often in the order of 20–33% of total operational costs when feedstocks are currently 'wastes', while processing costs usually are usually in the 70–80% range (Dien et al. 2003; Huber et al. 2006; Lin and Tanaka 2006; Solomon et al. 2007; Dale 2008). In the case of specific wastes, the share of feedstock costs in operational costs may even be lower. Joelsson and Gustavsson (2008) have, for instance, argued that a synthesis of transport biofuels based on the gasification of black liquor in the paper industry is competitive with mineral oil when the price of crude oil is at least US $40 per barrel. Black liquor is a co-product of paper that is relatively rich in lignin. The gas can be used for powering the paper plant and the production of transport biofuels such as methanol and dimethylether. In the case of crops grown as lignocellulosic feedstocks, the share of feedstock costs in operational costs may be higher than in the case of feedstocks that are currently wastes. Borgwardt (1999) considered lignocellulosic ethanol production with switchgrass or hybrid poplar as a feedstock and found that the feedstock cost was nearly 60% of operational costs.

Also, whether the current low (zero or even negative) costs of wastes and the relatively low costs of other lignocellulosic feedstocks can be maintained when they turn out to be good feedstocks for transport biofuel production is very doubtful. Indeed, in the long term, it seems likely that in this case, biofuels will follow the cost of competing fossil fuels, when corrected for differences in energy content (Naylor et al. 2007). When lignocellulosic biofuels turn out to be competitive, this may offer scope for substantial prices to be paid for what is currently considered a waste.

Still, it has been argued that as there are many sources of lignocellulose, it may well be that the price of feedstocks will be more stable than in case of starch or oil crops. This, however, is not necessarily relevant for production units turning out lignocellulosic biofuels. These may well restrict themselves to a limited range of feedstocks. Both in the case of enzymatic production and in the case of gasification, one would expect that production units, as they will be built in the near future, would be fit for a limited part of the broad range of lignocellulosic materials (e.g. Nathan 2007; Hayes 2008; Olofsson et al. 2008). On the other hand, it may well be that further technological development may allow for the use of broader ranges of lignocellulosic feedstocks.

Capital costs for converting lignocellulosic biomass into biofuel will be much higher than the capital costs for, for example, starch-based ethanol (Nathan 2007; Rotman 2008). Also, the operational costs of current enzymatic ways to produce lignocellulosic transport fuels are relatively high, even when currently available options for cost cutting and increasing the expected credit for co-products are implemented. For instance, in the case of enzymatic conversion, such costs are estimated to be greater than US $0.60 per litre (Sassner et al. 2008), whereas the 2006 costs for ethanol from Brazilian sugar cane were US $0.28–0.31 per litre (see Fig. 1.3). So, much reduced operating costs and increased yields would seem essential to the long-time financial viability of biochemical conversion of lignocellulose into ethanol. Overcoming the recalcitrance of cellulosic biomass, lower pre-treatment costs and lower costs of enzymatic conversions are priorities in this respect (Wyman 2007). Whether further research will indeed lead to much lower costs is an open question.

As to the prospects for future cost reduction of non-hydrolytic/fermentative ways to convert lignocellulosic biomass into transport fuels, the following may be noted. Some of the processes proposed for converting lignocellulose into transport fuels, such as the processes to convert synthesis gas into transport fuels, have been well researched and developed (Huber et al. 2006; Haryanto et al. 2007). However, gasification of biomass has only been subject to limited research, and it would seem that much can be done to optimize gasification of the wide range of feedstocks available (Nathan 2007; Wang et al. 2008b). In the field of gasification, there also would seem to be scope for cost reduction linked to technological developments such as membrane separation, supercritical water gasification and better control technology for tar, char and ashes (Han and Kim 2008; Haryanto et al. 2007; Wang et al. 2008b). The production of methane by anaerobic conversion of biomass is a well-developed technology, but scope for cost reduction and improvement of efficiency in the case of the conversion of lignocellulosic biomass to $CH_4$ may still be substantial (Bagi et al. 2007; Börjesson and Mattiasson 2008; Rodriguez et al. 2008).

If crop prices remain high, it may well be that, while excluding external costs, prices for many road transport biofuels may remain higher than fossil fuels in the near future. The higher cost of biofuels in the past has led to government policies favouring the application of biofuels. For the long-term viability of transport biofuels, however, it would seem unlikely that they can be more expensive than competitive fossil fuels. This may have a strong selective effect on production processes and producer countries.

## 1.8 Key Issues and the Rest of This Book

The growth of biofuel production and consumption for automotive transport is now the subject of a lively debate. This debate has led to revisions of transport biofuel policy in a number of countries, and it is likely that this debate will have further impacts on the development of biofuel production. Major items in this debate are the following.

### Energy Security

To the extent that transport biofuels are advocated to provide for energy security, it has been stated that their potential may be very limited. Eaves and Eaves (2007) have argued that devoting 100% of US corn to ethanol, while correcting for fossil fuel inputs, would displace 3.5% of gasoline consumption, 'only slightly more than the displacement that would follow from properly inflated tires'. Moreover, they have pointed out that historical US corn yields have shown considerable volatility, with corn yields about once every 20 years more than 30% lower than on average, which is not conducive to national energy security (Eaves and Eaves 2007). On the other hand, it has been argued that worldwide biofuels may end the dependence of transport on mineral oil, or even all fossil fuels. As pointed out above, de Vries et al. (2007) suggest that by 2050, up to about 300 EJ ($= 300 \times 10^{18}$ J) of liquid biofuels may be produced worldwide, mainly on abandoned agricultural soils, which would, as pointed out in Sect. 1.1, in all probability be sufficient to power transport.

### Food Prices and Food Security

The consequences of the increased use of transport biofuels for food prices and food security (access to affordable, adequate food supplies) have been other major topics in the debate on biofuels. In 2007, the rapid expansion of biofuels production contributed to increased prices for cereals and oilseeds (OECD-FAO 2007; Renewable Fuels Agency 2008). This effect of the growth in transport biofuel production on food prices has not gone unnoticed in society. Late in 2006, the Chinese government halted the expansion of corn-based ethanol production (Koizumi and Ohga 2007). At the beginning of 2007, Mexico was confronted by a tortilla crisis, including protests of poor people against rising prices for tortillas, which are made from corn. The Mexican government was forced to change its fiscal policy. Argentina, which has substantial soybean-based biodiesel production, raised its export taxes on soybeans by 4% to provide subsidies to lower the cost of soybean flour to livestock producers (OECD-FAO 2007), which in turn sparked angry farmer protests. In December 2007, the government of South Africa banned growing corn for biofuel to counteract price rises. 2008 saw widespread food rioting in Asia, Africa and South America. Several predictions suggest that a further rapid expansion of transport biofuel production will lead to (further) rises in food prices (Naylor et al. 2007; Eickhout et al. 2008) and that these rises may lead to an increased insufficiency of food for the world's poorest people that currently spend 50–80% of their total household income on food (Naylor et al. 2007; Runge and Senauer 2007; Daschle et al. 2007; Renewable Fuels Agency 2008). This has led to a coining of the slogan 'transport biofuels for the rich and hunger for the poor'.

### Environmental Concerns

In the debate on the future of transport biofuels, environmental matters have also become important. Slowing down climate change is often mentioned as an important reason for expanding the use of biofuels, as $CO_2$ that is released on burning this biomass is supposed to be rapidly sequestered again by re-growth of biomass. However, this is not the whole story. Because fossil fuels are used for powering biofuel production, biofuel production may be associated with the emission of greenhouse gases other than $CO_2$, such as $N_2O$ and $CH_4$, and biofuel production can be associated with changes in the carbon content of ecosystems. Thus, a lively discussion has originated on whether promoting transport biofuels does indeed slow down climate change. And there is a longstanding discussion as to whether the overall environmental impacts of biofuels are positive (e.g. Healy 1994), which has focused on the impacts of agricultural chemicals used in biomass production and water use. Also, the impact of transport biofuel production on nature has emerged as an issue. In particular, the importation by industrialized countries of palm oil from Southeast Asia, of biofuels from South America and the cutting of tropical forests for the establishment of biofuel plantations in Africa have sparked a debate on the impact of transport biofuels on living nature. In turn, environmental concerns contribute to the emergence of regulations and certification schemes that aim to address such concerns (Mathews 2008b; van Dam et al. 2008).

### Social Concerns

Social concerns have been raised, too, in the context of expanding transport biofuel production. These relate to land tenure, especially by native and small farmers confronted with expanding large-scale cropping of biofuel feedstocks, to the fate of such farmers at the hands of oppressive governments favouring large-scale biofuel projects, to labour relations, to working conditions and to the exploitation of child and migrant labour (Cooke 2002; Nicholls and Campos 2007; Smeets et al. 2008). Such social concerns have, for example, been raised about Burma, Malaysia, Indonesia, Brazil, Colombia and parts of Africa, in the context of ethanol production from sugar cane and palm oil and *Jatropha* oil production (Oxfam 2007; Ethnic Community Development Forum 2008; Gross 2008; Mayer 2008; Smeets et al. 2008). Social concerns contribute to the emergence of regulations and certification schemes that aim to address such concerns (Mathews 2008b; van Dam et al. 2008).

### This Book

The next chapters of this book will deal with matters that have a bearing on the debate about the future of biofuels with a main focus on environmental issues. Chapter 2 deals with cumulative fossil energy demand and solar energy conversion efficiencies of transport biofuels and of other ways to convert solar radiation into usable energy. These are important for the potential to displace fossil fuels and the area required for biofuel supply. Chapter 3 takes a look at the use of non-energy resources for transport biofuel production, such as water, plant nutrients and fertile soils. Chapter 4 considers the emissions linked to the life cycle of biofuels. Chapter 5 discusses the impact of transport biofuel production on living nature. Chapter 6 looks at the future of transport biofuels in view of the previous chapters and tries to answer questions that are frequently asked about biofuels.

# References


Abdullah AZ, Razali N, Mootabadi H, Salamatinia B (2007) Critical technical areas for future improvement in biodiesel technologies. Environ Res Lett 2:034001

Acharya V, Young BR (2008) A review of the potential of bio-ethanol in New Zealand. B Sci Technology & Society 28:143–148

Achten WMJ, Verschot L, Franken VJ, Mathijs E, Singh VP, Aerts R, Muys B (2009) Jatropha biodiesel production and use. Biomass Bioenerg in press

Agarwal D, Agarwal AK (2007) Performance and emissions characteristics of Jatropha oil (preheated and blends) in a direct injection compression ignition engine. Appl Therm Eng 27: 2314–2323

Agnolucci P (2007) Hydrogen infrastructure for the transport sector. Int J Hydrogen Energ 32:3526–3544

Ahman M, Nilsson LJ (2008) Path dependency and the future of advanced vehicles and biofuels. Utilities Policy 16:80–8

Alimahmoodi M, Mulligan CN (2008) Anaerobic bioconversion of carbon dioxide to biogas in an upflow anaerobic sludge blanket reactor. J Air Waste Manag Assoc 58:95–103

Amigun B, Sigamoney R, von Blottnitz H (2008) Commercialisation of biofuel industry in Africa: a review. Renew Sust Energ Rev 12:690–711

Anderson WF, Akin DE (2008) Structural and chemical properties of grass lignocelluloses related to conversion for biofuels. J Ind Microbiol Biotechnol 35:355–366

Andersson E, Harvey S (2006) System analysis of hydrogen production from gasified black liquor. Energy 31:3426–3434

Angenent LT (2007) Energy biotechnology: beyond the general lignocellulose-to-ethanol pathway. Curr Opin Biotechnol 18:191–192

Anonymous (2006) DME – the diesel alternative which is easy to dismiss, but hard to beat. Hydrocarbon Processing Magazine March:29–30

Antoni D, Zverlov VV, Schwarz WH (2007) Biofuels from microbes. Appl Microbiol Biotechnol 77:23–35

Archer D (2005) Fate of fossil fuel $CO_2$ in geologic time. J Geophys Res 110:C09S05

Arcoumanis C, Bae C, Crookes R, Kinoshita E (2008) The potential of di-methyl ether (DME) as an alternative fuel for compression-ignition engines: a review. Fuel 87:1014–1030

Arifeen N, Wang R, Kookos I, Webb C, Koutinas AA (2007) Optimization and cost estimation of novel wheat biorefining for continuous production of fermentation feedstock. Biotechnol Progr 23:872–880

Ashley S (2008) Cellulose success. Sci Am 298(4):32–33

Astbury GR, Hawksworth SJ (2007) Spontaneous ignition of hydrogen leaks: a review of postulated mechanisms. Int J Hydrogen Energ 32:2178–2185

Atsumi S, Hanai T, Liao JC (2008) Non-fermentative pathways for synthesis of branched-chain higher alcohols as biofuels. Nature 451:86–89

Bagajewicz M, Sujo D, Martinez D, Savelski M (2007) Driving without petroleum? A comparative guide to biofuels, gas-to-liquids and coal-to-liquids as fuels for transportation. Energy Charter Secretariat, Brussels

Bagi Z, Ács N, Bálint B, Horváth L, Dobó K, Perei KR, Rákhely G, Kovács KL (2007) Biotechnological intensification of biogas production. Appl Microbiol Biotechnol 76:473–482

Ballesteros M, Oliva JM, Negro MJ, Manzanares P, Ballesteros I (2004) Ethanol from lignocellulosic materials by a simultaneous saccharification and fermentation process (SFS) with *Kluyveromyces marxianus* CECT 10875. Process Biochem 39:1843–1848

Banerjee A, Sharma R, Chisti Y, Banerjee UC (2002) *Botryococcus braunii*: a renewable source of hydrocarbons and other chemicals. Crit Rev Biotechnol 22:245–279

Barelli L, Bidini G, Gallorini F, Servili S (2008) Hydrogen production through sorption-enhanced steam methane reforming and membrane technology: a review. Energy 33:554–570

Barlaz MA (2006) Forest products decomposition in municipal solid waste landfills. Waste Manage 26:321–333

Barrett D (2007) Experts address the question: given its relatively high cost, is renewable energy the answer for SIDS? Nat Resour Forum 31:162

Bastianoni S, Coppola F, Tiezzi E, Colacevich A, Borghini F, Focardi S (2008) Biofuel potential production from the Orbetello lagoon macroalgae: a comparison with sunflower feedstock. Biomass Bioenerg 32:619–628

Ben-Amotz A, Sussman I, Avron M (1982) Glycerol production by *Dunaliella*. Experientia 38: 49–52

Bentley RW, Mannan SA, Wheeler SJ (2007) Assessing the date of the global oil peak: the need to use 2P reserves. Energ Policy 35:6364–6382

Berggren M, Ljunggren E, Johnsson F (2008) Biomass co-firing potentials for electricity generation in Poland – matching supply and co-firing opportunities. Biomass and Bioenerg 32: 865–879

Berndes G, Hansson J (2007) Bioenergy expansion in the EU: cost-effective climate change mitigation, employment creation and reduced dependency on imported fuels. Energ Policy 35: 5965–5979

Bideaux C, Alfenore S, Cameleyre X, Molina-Jouve C, Uribelarrea J, Guillouet SE (2006) Minimization of glycerol production during the high-performance fed-batch ethanolic fermentation process in *Saccharomyces cerevisiae*, using a metabolic model as a prediction tool. Appl Exp Microbiol 72:2134–2140

Blanch HW, Adams PD, Andrews-Cramer KM, Frommer WB, Simmons BA, Keasling JD (2008) Addressing the need for alternative transportation fuels: the Joint BioEnergy Institute. ACS Chem Biol 3(1):17–20

Boateng AA, Anderson WF, Phillips JG (2007) Bermudagrass for biofuels: effect of two genotypes on pyrolysis product yield. Energ Fuel 21:1183–1187

Bocher BT, Agler MT, Garcia ML, Beers AR, Angenent LT (2008) Anaerobic digestion of secondary residuals from an anaerobic bioreactor at a brewery to enhance bioenergy generation. J Ind Microbiol Biotechnol 35:321–329

Boer R, Wasrin UR, Perdinan, Hendri, Dasanto BD, Makundi W, Hero J, Ridwan M, Masripatin N (2007) Assessment of carbon leakage in multiple carbon-sink projects: a case study in Jambi Province, Indonesia. Mitigation Adaptation Strategies Global Change 12:1169–1188

Borgwardt RH (1999) Transportation fuel from cellulosic biomass: a comparative assessment of ethanol and methanol options. P I Mech Eng A–J Pow 213:399–407

Börjesson P, Mattiasson B (2008) Biogas as a resource-efficient vehicle fuel. Trends Biotechnol 26:7–13

Bridgwater AV, Meier D, Radlein D (1999) An overview of fast pyrolysis of biomass. Org Geochem 30:1479–1493

Bryner M (2007a) DOE to make up to $385-million investment in six biorefinery projects. Chem Week March 7:11

Bryner M (2007b) Alternative fuels. Chem Week December 13/20:18–20

Bungay HR (2004) Confessions of a bioenergy advocate. Trends Biotechnol 22:67–71

Buranov AU, Mazza G (2008) Lignin in straw of herbaceous crops. Ind Crop Prod 28:237–259

Buschmann AH, Correa JA, Westermeier R, Hernández-González MdC, Norambuena R (2001) Red algal farming in Chile: a review. Aquaculture 194:203–220

Bush GW (2006) State of the Union address. http://www.whitehouse.gov/stateoftheunion/2006/

Canakci M, Sanli H (2008) Biodiesel production from various feedstocks and their effects on the fuel properties. J Ind Microbiol Biotechnol 35:431–441

Cantrell KB, Ducey T, Ro KS, Hunt PG (2008) Livestock waste-to-bioenergy generation opportunities. Bioresour Technol 99:7941–7953

Carcaillet C, Talon B (2001) Soil carbon sequestration by Holocene fires inferred from soil charcoal in the dry French Alps. Arct Antarct Alp Res 33:282–288

Carvalho CR, Clarindo WR, Praça MM, Araújo FS, Carels N (2008) Genome size, base composition and karyotype of *Jatropha curcas* L., an important biofuel plant. Plant Sci 174:613–617

Cascone R (2007) Biofuels: what is beyond ethanol and biodiesel? Hydrocarbon Processing Magazine September:95–109

Chalk SG, Miller JF (2006) Key challenges and recent progress in batteries, fuel cells, and hydrogen storage for clean energy systems. J Power Sources 159:73–80

Chang MCY (2007) Harnessing energy from plant biomass. Curr Opin Chem Biol 11:677–684

Chapple C, Ladisch M, Meilan R (2007) Loosening lignin's grip on biofuel production. Nat Biotechnol 25:746–748

Cheng CH, Cheung CS, Chan TL, Lee SC, Yao CD, Tsang KS (2008) Comparison of emissions of a direct injection diesel engine operating on biodiesel with emulsified and fumigated methanol. Fuel 87:1870–1879

Chiao J, Sun ZH (2007) History of the acetone-butanol-ethanol fermentation industry in China: development of continuous production technology. J Mol Microbiol Biotechnol 13:12–14

China Chemical Reporter (2007) Develop bio-diesel and accelerate energy substitution. July 6:7

Chisti Y (2007) Biodiesel from microalgae. Biotechnol Adv 25:294–306

Chopin T, Buschmann AH, Halling C, Troell M, Kautsky N, Neori A, Kraemer GP, Zertuche-González JA, Yarish C, Neefus C (2001) Integrating seaweeds into marine aquaculture systems: a key towards sustainability. J Phycol 37:975–986

Claassen PAM, van Lier JB, Lopez Contreras AM, van Niel EWJ, Sijtsma L, Stams AJM, de Vries SS, Weusthuis RA (1999) Utilisation of biomass for the supply of energy carriers. Appl Microbiol Biotechnol 52:741–755

Cloin J (2007) Coconut oil as a fuel in the Pacific Islands. Nat Resour Forum 31:119–127

Colella WG, Jacobson MZ, Golden DM (2005) Switching to a U.S. hydrogen fuel cell vehicle fleet: the resultant change in emissions, energy use, and greenhouse gases. J Power Sources 150:150–181

Cooke FM (2002) Vulnerability, control and oil palm in Sarawak: globalization and a new era? Dev Change 33:189–211

Coombs A (2007) Glycerin bioprocessing goes green. Nat Biotechnol 25:953–954

Critchley AT, Ohno M, Largo DB (eds) (2006) World seaweed resources: an authoritative reference system. ETI Information Services, Wokingham (UK)

da Costa MM, Cohen C, Schaeffer R (2007) Social features of energy production and use in Brazil: goals for a sustainable energy future. Nat Resour Forum 31:11–20

Daggett DL, Hendricks RC, Walther R, Corporan E (2007) Alternate fuels for use in commercial aircraft. Boeing, Seattle

Dale B (2008) Biofuels: thinking clearly about the issues. J Agric Food Chem 56:3885–3891

Daschle T, Runge CF, Senauer B (2007) Debating the tradeoffs of corn-based ethanol: myth versus reality. Foreign Affairs September/October

de la Rue du Can S, Price L (2008) Sectoral trends in global energy use and greenhouse gas emissions. Energ Policy 36:1386–1403

Delucchi MA, Lipman TE (2001) An analysis of the retail and lifecycle cost of battery-powered electric vehicles. Transport Res D–Tr E 6:371–404

Delucchi MA, Murphy JJ (2008) US military expenditures to protect the use of Persian Gulf oil for motor vehicles. Energ Policy 36:2253–2264

Demain AL, Newcomb M, Wu JHD (2005) Cellulase, Clostridia, and ethanol. Microbiol Mol Biol R 69:124–154

Demiral J, Şensöz S (2006) Fixed-bed pyrolysis of hazelnut (*Corylus Avellana* L.) bagasse: influence of pyrolysis parameters on product yields. Energ Source Part A 28:1149–1158

Demirbaş A (2001) Biomass resource facilities and biomass conversion processing for fuels and chemicals. Energ Convers Manage 42:1357–1378

Demirbaş A, Kara H (2006) New options for conversion of vegetable oils to alternative fuels. Energ Source 28:619–626(8)

Demirbaş A (2007) Importance of biodiesel as transportation fuel. Energ Policy 35:4661–4670

Demirbaş A (2008) Producing bio-oil from olive cake by fast pyrolysis. Energ Source Part A 30:38–44

Desvaux M (2005) *Clostridium cellulolyticum*: model organism of mesophilic cellulolytic clostridia. FEMS Microbiol Rev 29:741–764

de Vries BJM, van Vuuren DP, Hoogwijk MM (2007) Renewable energy sources: their global potential for the first-half of the 21st century at a global level: an integrated approach. Energ Policy 35:2590–2610

Di Blasi C (2008) Modeling chemical and physical processes of wood and biomass pyrolysis. Prog Energ Combust 34:47–90

Dien BS, Cotta MA, Jeffries TW (2003) Bacteria engineered for fuel ethanol production: current status. Appl Microbiol Biotechnol 63:248–266

Dietenberger MA, Anderson M (2007) Vision of the U.S. biofuel future: a case for hydrogen-enriched biomass gasification. Ind Eng Chem Res 46:8863–8874

Di Lucia L, Nilsson LJ (2007) Transport biofuels in the European Union: the state of play. Transp Policy 14:533–543

Dimitri C, Effland A (2007) Fueling the automobile: an economic exploration of early adoption of gasoline over ethanol. J Agric Food Ind Organ 5(2):article 11. http://www.bepress.com/jafio/vol5/iss2/art11

Dismukes GC, Carrieri D, Bennette N, Ananyev GM, Posewitz MC (2008) Aquatic phototrophs: efficient alternatives to land-based crops for biofuels. Curr Opin Biotechnol 19:235–240

Dobele G, Urbanovich I, Volpert A, Kampars V, Samulis E (2007) Fast pyrolysis – effect of wood drying on the yield and properties of bio-oil. BioResources 2:699–706

Doering C (2008) US lawmakers urge scaling back biofuels mandate. Planet Ark May 7. http://www.planetark.com/dailynewsstory.cfm/newsid/48252/story.htm

Dondero L, Goldemberg J (2005) Environmental implications of converting light gas vehicles: the Brazilian experience. Energ Policy 33:1703–1708

Dumas C, Basseguy R, Bergel A (2008) Microbial electrocatalysis with *Geobacter sulfurreducens* biofilm on stainless steel cathodes. Electrochim Acta 53:2494–2500

Dupain X, Costa DJ, Schaverien CJ, Makkee M, Moulijn JA (2007) Cracking of a rapeseed vegetable oil under realistic FCC conditions. Appl Catal B–Environ 72:44–61

Dürre P (2008) Fermentative butanol production: bulk chemical and biofuel. Ann NY Acad Sci 1125:353–362

Eaves J, Eaves S (2007) Renewable corn-ethanol and energy security. Energ Policy 35:5958–5963

Eckmeier E, Rösch M, Ehrmann O, Schmidt MWI, Schier W, Gerlach R (2007) Conversion of biomass to charcoal and the carbon mass balance from a slash-and-burn experiment in a temperate deciduous forest. Holocene 17:539–542

Eickhout B, van den Born GJ, Notenboom J, van Oorschot M, Ros JPM, van Vuuren DP, Westhoek HJ (2008) Local and global consequences of the EU renewable directive for biofuels. Milieu en Natuur Planbureau Bilthoven. http://www.mnp.nl

Eijsink VGH, Vaaje-Kolstad G, Varum KM, Horn SJ (2008) Towards new enzymes for biofuels: lessons from chitinase research. Trends Biotechnol 26:228–235

El Bassam N (1998) $C_3$ and $C_4$ plant species as energy sources and their potential impact on environment and climate. Renew Energ 15:205–210

Esler D (2007) Alternative fuels for jet engines. Business and Commercial Aviation 101:(3)01914624

Ethnic Community Development Forum (2008) Biofuel by decree: unmasking Burma's bio-energy fiasco. http://www.terraper.org/file_upload/BiofuelbyDecree.pdf

EUCAR, CONCAWE, European Commission JRC (2007) Well-to-wheels analysis of future automotive fuels and powertrains in the European context, vers 2c. http://ies.jrc.ec.europa.eu/uploads/media/WTW_Report_010307.pdf

European Union (2008) Directive on the promotion of the use of renewable energy sources. Brussels

Ewers RM, Rodrigues ASL (2008) Estimates of reserve effectiveness are confounded by leakage. Trends Ecol Evol 23:113–116

Ezeji TC, Qureshi N, Blaschek HP (2007) Bioproduction of butanol from biomass: from genes to bioreactors. Curr Opin Biotechnol 18:220–227

Fahmi R, Bridgwater AV, Donnison I, Yates N, Jones JM (2008) The effect of lignin and inorganic species in biomass on pyrolysis oil yields, quality and stability. Fuel 87:1230–1240
Fairless D (2007) Biofuel: the little shrub that could – maybe. Nature 449:652–655
Felder R, Dones R (2007) Evaluation of ecological impacts of synthetic natural gas from wood used in current heating and car systems. Biomass Bioenerg 31:403–415
Fernando S, Hanna M (2004) Development of a novel biofuel blend using ethanol-biodiesel-diesel microemulsions: EB-diesel. Energ Fuel 18:1695–1703
Fernando S, Adhikari S, Kota K, Bandi R (2007) Glycerol based automotive fuels from future biorefineries. Fuel 86:2806–2809
Ferreira-Aparicio P, Benito MJ, Sanz JL (2005) New trends in reforming technologies: from hydrogen industrial plants to multifuel microreformers. Catal Rev 47:491–588
Florin NH, Harris AT (2008) Enhanced hydrogen production from biomass with in situ carbon dioxide capture using calcium oxide sorbents. Chem Eng Sci 63:287–316
Fortman JL, Chhabra S, Mukhopadhyay A, Chou H, Lee TS, Steen E, Keasling JD (2008) Biofuel alternatives to ethanol: pumping the microbial well. Trends Biotechnol 26:375–381
Fowles M (2007) Black carbon sequestration as an alternative to bioenergy. Biomass Bioenerg 31:426–432
Francis G, Edinger R, Becker K (2005) A concept for simultaneous wasteland reclamation, fuel production, and socio-economic development in degraded areas in India: need, potential and perspectives of *Jatropha* plantations. Nat Resour Forum 29:12–24
Frederick WJ Jr, Lien SJ, Courchene CE, DeMartini NA, Ragauskas AJ, Iisa K (2008) Production of ethanol from carbohydrates from loblolly pine: a technical and economic assessment. Bioresour Technol 99:5051–5057
GAO (United States Government Accountability Office) (2007) Crude oil: uncertainty about future oil supply makes it important to develop a strategy for addressing a peak and decline in oil production. GAO-07-283. http://www.gao.gov/new.items/d07283.pdf
Gassmann A, Cock MJW, Shaw R, Evans HC (2006) The potential for biological control of invasive alien aquatic weeds in Europe: a review. Hydrobiologia 570:217–222
Gibbs HK, Johnston M, Foley JA, Holloway T, Monfreda C, Ramankutty N, Zaks D (2008) Carbon payback times for crop-based biofuel expansion in the tropics: the effects of changing yield and technology. Environ Res Lett 3:034001
Gjoen H, Hard M (2002) Cultural politics in action: developing user scripts in relation to the electric vehicle. Sci Technol Hum Val 27:262–281
Goldemberg J, Coelho ST, Nastari PM, Lucon O (2004) Ethanol learning curve – the Brazilian experience. Biomass Bioenerg 26:301–304
Gomez LD, Steele-King CG, McQueen-Mason SJ (2008) Sustainable liquid biofuels from biomass: the writing's on the walls. New Phytol 178:473–485
Gough CM, Vogel CS, Schmid HP, Curtis PS (2008) Controls on annual forest carbon storage: lessons from the past and predictions for the future. BioScience 58:609–622
Goyal HB, Seal D, Saxena RC (2008) Bio-fuels from thermochemical conversion of renewable resources: a review. Renew Sust Energ Rev 12:504–517
Gray KA, Zhao L, Emptage M (2006) Bioethanol. Curr Opin Chem Biol 10:141–146
Grimm C (1999) Evaluation of damage to physic nut (*Jatropha curcas*) by true bugs. Entomol Exp Appl 92:127–136
Gross M (2008) Not in our backyard. Curr Biol 18:R227–R228
Güllü D, Demirbaş A (2001) Biomass to methanol via pyrolysis process. Energ Convers Manage 42:1349–1356
Gunnarsson CC, Petersen CM (2007) Water hyacinths as a resource in agriculture and energy production: a literature review. Waste Manage 27:117–129
Guschina IA, Harwood JL (2006) Lipids and lipid metabolism in eukaryotic algae. Prog Lipid Res 45:160–186
Gutierrez NA, Maddox IS, Schuster KC, Swoboda H, Gapes JR (1998) Strain comparison and medium preparation for the acetone-butanol-ethanol (ABE) fermentation process using a substrate of potato. Bioresour Technol 66:263–265

Hahn JJ, Ghirardi ML, Jacoby WA (2007) Immobilized algal cells used for hydrogen production. Biochem Eng J 37:75–79
Hahn-Hägerdal B, Karhumaa K, Fonseca C, Spencer-Martins I, Gorwa-Grauslund MF (2007) Towards industrial pentose-fermenting yeast strains. Appl Microbiol Biotechnol 74:937–953
Hall DO (1982) Solar energy through biology: fuels from biomass. Experientia 38:3–10
Hall KR (2005) A new gas to liquids (GTL) or gas to ethylene (GTE) technology. Catal Today 106:243–246
Hallenbeck PC, Benemann JR (2002) Biological hydrogen production; fundamentals and limiting processes. Int J Hydrogen Energ 27:1185–1193
Hammerschlag R (2006) Ethanol's energy return on investment: a survey of the literature 1990–present. Environ Sci Technol 40:1744–1750
Hammond GP, Kallu S, McManus MC (2008) Development of biofuels for the UK automotive market. Appl Energ 85:506–515
Han J, Kim H (2008) The reduction and control technology of tar during biomass gasification/pyrolysis: an overview. Renew Sust Energ Rev 12:397–416
Hankamer B, Lehr F, Rupprecht J, Mussgnug JH, Posten C, Kruse O (2007) Photosynthetic biomass and $H_2$ production by green algae: from bioengineering to bioreactor scale-up. Physiol Plantarum 131:10–21
Hansen J, Sato M, Ruedy R, Lacis A, Oinas V (2000) Global warming in the twenty-first century: an alternative scenario. P Natl Acad Sci USA 97:9875–9880
Hansen J, Sato M, Kharecha P, Beerling D, Berner R, Masson-Delmotte V, Pagani M, Raymo M, Royer DL, Zachos JC (2008) Target atmospheric $CO_2$: where should humanity aim? Open Atmos Sci J in press
Harding KG, Dennis JS, von Blottnitz H, Harrison STL (2008) A life-cycle comparison between inorganic and biological catalysis for the production of biodiesel. J Clean Prod 16:1368–1378
Haryanto A, Fernando S, Adhikari S (2007) Ultrahigh temperature water gas shift catalysts to increase hydrogen yield from biomass gasification. Catal Today 129:269–274
Hayes DJ (2008) An examination of biorefining processes, catalysts and challenges. Catal Today in press
Healy SA (1994) The recent European biofuel debate as a case study in the politics of renewable energy. Renew Energ 5:875–877
Heaton EA, Flavell RB, Mascia PN, Thomas SR, Dohleman FG, Long SP (2008) Herbaceous energy crop development: recent progress and future prospects. Curr Opin Biotechnol 19: 202–209
Heiman MK, Solomon BD (2007) Fueling US transportation: the hydrogen economy and its alternatives. Environment 49(8):10–25
Heinimö J (2008) Methodological aspects on international biofuels trade: international streams and trade of solid and liquid biofuels in Finland. Biomass Bioenerg 32:702–716
Helle SS, Lin T, Duff SJB (2008) Optimization of spent sulfite liquor fermentation. Enzyme Microb Technol 42:259–264
Henstra AM, Sipma J, Rinzema A, Stams AJM (2007) Microbiology of synthesis gas fermentation for biofuel production. Curr Opin Biotechnol 18:200–206
Hirano A, Hon-Nami K, Kunito S, Hada M, Ogushi Y (1998) Temperature effect on continuous gasification of microalgal biomass: theoretical yield of methanol production and its energy balance. Catal Today 45:399–404
Holmgren J, Gosling C, Marinangeli R, Marker T, Faraci G, Perego C (2007) New developments in renewable fuels offer more choices. Hydrocarbon Processing Magazine September:67–71
Horn SJ, Aasen IM, Østgaard K (2000) Ethanol production from seaweed extract. J Ind Microbiol Biotechnol 25:249–254
Høyer KG (2008) The history of alternative fuels in transportation: the case of electric and hybrid cars. Utilities Policy 16:63–71
Huang H, Ramaswamy S, Tschirner UW, Ramarao BV (2008) A review of separation technologies in current and future biorefineries. Sep Purif Technol 62:1–21

Huber GW, Iborra S, Corma A (2006) Synthesis of transportation fuels from biomass: chemistry, catalysts, and engineering. Chem Rev 106:4044–4098

Huntley ME, Redalje DG (2007) $CO_2$ mitigation and renewable oil from photosynthetic microbes: a new appraisal. Mitigation Adaptation Strategies Global Change 12:573–608

Hussain MM, Dincer I, Li X (2007) A preliminary life cycle assessment of PEM fuel cell powered automobiles. Appl Therm Eng 27:2294–2299

Janssen A, Lienin SF, Gassmann F, Wokaun A (2006) Model aided policy development for the market penetration of natural gas vehicles in Switzerland. Transport Res A–Pol 40:316–333

Joelianingsih, Maeda H, Hagiwara S, Nabetani H, Sagara Y, Soerawidjaya TH, Tambunan AH, Abdullah K (2008) Biodiesel fuels from palm oil via the non-catalytic transesterification in a bubble column reactor at atmospheric temperature: a kinetic study. Renew Energ 33: 1629–1636

Joelsson JM, Gustavsson L (2008) $CO_2$ emission and oil use reduction through black liquor gasification and energy efficiency in pulp and paper industry. Resour Conserv Recy 52:747–763

Johansson B (1999) The economy of alternative fuels when including the cost of air pollution. Transport Res D–Tr E 4:91–108

Johnston M, Holloway T (2007) A global comparison of national biodiesel production potentials. Environ Sci Technol 41:7967–7973

Joint I, Henriksen P, Garde K, Riemann B (2002) Primary production, nutrient assimilation and microzooplankton grazing along a hypersaline gradient. FEMS Microbiol Ecol 39:245–257

Jones PR (2008) Improving fermentative biomass-derived $H_2$-production by engineering microbial metabolism. Int J Hydrogen Energ 33:5122–5130

Jones DT, Woods DR (1986) Acetone-butanol fermentation revisited. Microbiol Mol Biol Rev 50:484–524

Junginger M, de Wit M, Sikkema R, Faaij A (2008) International bioenergy trade in the Netherlands. Biomass Bioenerg 32:672–687

Kadam KL, Chin CY, Brown LW (2008) Flexible biorefinery for producing fermentation sugars, lignin and pulp from corn stover. J Ind Microbiol Biotechnol 35:331–341

Kalita D (2008) Hydrocarbon plant – new source of energy for future. Renew Sust Energ Rev 12:455–471

Kalscheuer R, Stölting T, Steinbüchel A (2006) Microdiesel: *Escherichia coli* engineered for fuel production. Microbiology 152:2529–2536

Kamimura A, Sauer IL (2008) The effect of flex fuel vehicles in the Brazilian light road transportation. Energ Policy 36:1574–1576

Kamm B, Kamm M (2004) Principles of biorefineries. Appl Microbiol Biotechnol 64:137–145

Kamm J (2004) A new class of plants for a biofuel feedstock energy crop. Appl Biochem Biotechnol 113:55–70

Kaufmann RK, Shiers LD (2008) Alternatives to conventional crude oil: when, how quickly, and market driven? Ecol Econ 67:405–411

Kegl B (2008) Effects of biodiesel on emissions of a bus diesel engine. Bioresour Technol 99: 863–873

Kerschbaum S, Rinke G, Schubert K (2008) Winterization of biodiesel by micro process engineering. Fuel 87:2590–2597

Kheshgi HS, Prince RC, Marland G (2000) The potential of biomass fuels in the context of global climate change: focus on transportation fuels. Annu Rev Energ Env 25:199–244

Kheshgi HS, Prince RC (2005) Sequestration of fermentation $CO_2$ from ethanol production. Energy 30:1865–1871

Kirilenko AP, Sedjo RA (2007) Climate change impacts on forestry. P Natl Acad Sci USA 104:19697–19702

Kleerebezem R, van Loosdrecht MCM (2007) Mixed culture biotechnology for bioenergy production. Curr Opin Biotechnol 18:207–212

Kleiner K (2007) Civil aviation faces green challenge. Nature 448:120–121

Kleinert M, Barth T (2008) Towards a lignincellulosic biorefinery: direct one-step conversion of lignin to hydrogen-enriched biofuel. Energ Fuel 22:1371–1379

Knothe G (2001) Historical perspectives on vegetable oil-based diesel fuels. Inform 12:1103–1107
Koizumi T, Ohga K (2007) Biofuels policies in Asian countries: impact of the expanded biofuels programs on world agricultural markets. J Agric Food Ind Organ 5(2):article 8. http://www.bepress.com/jafio/vol5/iss2/art8/
Kondarides DI, Daskalaki VM, Patsoura A,Verykios XE (2008) Hydrogen production by photo-induced reforming of biomass components and derivatives at ambient conditions. Catal Lett 122:26–32
Körbitz W (1999) Biodiesel production in Europe and North America, an encouraging prospect. Renew Energ 16:1078–1083
Lache R, Galves D, Nolan P (2008) Electric cars: plugged in. Deutsche Bank Securities Inc., New York
Lachke A (2002) Biofuel from D-xylose – the second most abundant sugar. Resonance 7:50–58
Lamers P, McCormick K, Hilbert JA (2008) The emerging liquid biofuel market in Argentina: implications for domestic demand and international trade. Energ Policy 36:1479–1490
Lange JP (2007) Lignocellulose conversion: an introduction into chemistry, process and economics. Biofuel Bioprod Biorefining 1:39–48
Lapeña-Rey N, Mosquera J, Bataller E, Orti F, Dudfield C, Orsillo A (2008) Environmentally friendly power sources for aerospace applications. J Power Sources 181:353–362
Lehmann J, Gaunt J, Rondon M (2006) Bio-char sequestration in terrestrial ecosystems – a review. Mitigation Adaptation Strategies Global Change 11:395–419
Lehtomäki A, Huttunen S, Lehtinen TM, Rintala JA (2008) Anaerobic digestion of grass silage in batch leach bed processes for methane production. Bioresour Technol 99:3267–3278
Lewis K (1966) Symposium on bioelectrochemistry of microorganisms. IV. Biochemical fuel cells. Bacteriol Rev 30:101–113
Li C, Wang Q, Zhao ZK (2008) Acid in ionic liquid: an efficient system for hydrolysis of lignocellulose. Green Chem 10:177–182
Li X, Lu G, Guo Y, Guo Y, Wang Y, Zhang Z, Liu X, Wang Y (2007) A novel solid superbase of $Eu_2O_3/Al_2O_3$ and its catalytic performance for the transesterification of soybean oil to biodiesel. Catal Commun 8:1969–1972
Licht FO (2006) World Ethanol and Biofuels Report
Lin Y, Tanaka S (2006) Ethanol fermentation from biomass resources: current state and prospects. Appl Microbiol Biotechnol 69:627–642
Littlefield S, Nickens A (2005) Roadmap for the all-electric warship. Power 149(1):46–50
Liu Z, Wang G, Zhou B (2008) Effect of iron on growth and lipid accumulation in *Chlorella vulgaris*. Bioresour Technol 99:4717–4722
Lobell DB, Burke MB, Tebaldi C, Mastrandrea MD, Falcon WP, Naylor RL (2008) Prioritizing climate change adaptation needs for food security in 2030. Science 319:607–610
Long CJ, Whitlock C, Bartlein PJ (2007) Holocene vegetation and fire history of the Coast Range, western Orgeon, USA. Holocene 17:917–926
Lynd LR (1996) Overview and evaluation of fuel ethanol from cellulosic biomass: technology, economics, the environment, and policy. Ann Rev Energ Env 21:403–465
Lynd LR, Weimer PJ, van Zyl WH, Pretorius IS (2002) Microbial cellulose utilization: fundamentals and biotechnology. Microbiol Mol Biol Rev 66:506–577
Lynd LR, Laser MS, Bransby D, Dale BE, Davison B, Hamilton R, Himmel M, Keller M, McMillan JD, Sheehan J, Wyman CE (2008) How biotech can transform biofuels. Nat Biotechnol 26:169–172
Mabee W, Roy DN (2003) Modeling the role of papermill sludge in the organic carbon cycle of paper products. Environ Rev 11:1–16
Macedo IC, Seabra JEA, Silva JEAR (2008) Greenhouse gases emissions in the production and use of ethanol from sugarcane in Brazil: the 2005/2006 averages and a prediction for 2020. Biomass Bioenerg 32:582–595
MacLean HL, Lave LB (2003) Evaluating automobile fuel/propulsion system technologies. Prog Energ Combust 29:1–69

Malik A (2007) Environmental challenge *vis a vis* opportunity: the case of water hyacinth. Environ Int 33:122–138
Marchetti JM, Miguel VU, Errazu AF (2007) Possible methods for biodiesel production. Renew Sust Energ Rev 11:1300–1311
Markert F, Nielsen SK, Paulsen JL, Andersen V (2007) Safety aspects of future infrastructure scenarios with hydrogen refuelling stations. Int J Hydrogen Energ 32:2227–2234
Marques S, Alves L, Roseiro JC, Girio FM (2008) Conversion of recycled paper sludge to ethanol by SHF and SSF using *Pichia stipitis*. Biomass Bioenerg 32:400–406
Mathews JA (2008a) Carbon-negative biofuels. Energ Policy 36:940–945
Mathews JA (2008b) Towards a sustainably certifiable futures contract for biofuels. Energ Policy 36:1577–1583
Matheys J, van Autenboer W, Timmermans J, van Mierlo J, van den Bossche P, Maggetto G (2007) Influence of functional unit on the life cycle assessment of traction batteries. Int J Life Cycle Ass 12:191–196
Mayer J (2008) Borneo project: burning for biofuels. Earth Isl J March 22
McCann MC, Carpita NC (2008) Designing the deconstruction of plant cell walls. Curr Opin Plant Biol 11:314–320
McElroy AK (2007) Not so run of the mill. Biomass Magazine October. http://www.biomassmagazine.com/article.jsp?article_id=1297
Melaina MW (2007) Turn of the century refueling: a review of innovations in early gasoline refueling methods and analogies for hydrogen. Energ Policy 35:4919–4934
Melis A, Happe T (2001) Hydrogen production: green algae as a source of energy. Plant Physiol 127:740–748
Melis A, Melnicki MR (2006) Integrated biological hydrogen production. Int J Hydrogen Energ 31:1563–1573
Mikkonen S (2008) Second-generation renewable diesel offers advantages. Hydrocarbon Processing Magazine 87:63–68
Milne TA, Evans RJ, Nagle N (1990) Catalytic conversion of microalgae and vegetable oils to premium gasoline, with shape-selective zeolites. Biomass 21:219–232
Mom G (1997) Das 'Scheitern' des frühen Elektromobils (1895–1925). Versuch einer Neubewertung. Technikgeschichte 64:269–285
Mom GPA, Kirsch DA (2001) Technologies in tension: horses, electric trucks, and the motorization of American cities, 1900–1925. Technol Cult 42:489–518
Monfort J (2008) Despite obstacles, biofuels continue surge. World Watch 21(4):34–35
Monti A, Di Virgilio N, Venturi G (2008) Mineral composition and ash content of six major energy crops. Biomass Bioenerg 32:216–223
Morand P, Merceron M (2005) Microalgal population and sustainability. J Coastal Res 21: 1009–1020
Moreira JR (2006) Global biomass energy potential. Mitigation Adaptation Strategies Global Change 11:313–333
Müller-Hagedorn M, Bockhorn H (2007) Pyrolytic behaviour of different biomasses (angiosperms) (maize plants, straws, and wood) in low temperature pyrolysis. J Anal Appl Pyrol 79:136–146
Murphy JD, McCarthy K (2005) The optimal production of biogas for use as a transport fuel in Ireland. Renew Energ 30:2111–2127
Nath K, Das D (2004) Improvement of fermentative hydrogen production: various approaches. Appl Microbiol Biotechnol 65:520–529
Nathan S (2007) Field of dreams? Engineer July–August:20–24
Naylor RL, Liska AJ, Burke MB, Falcon WP, Gaskell JC, Rozelle SD, Cassman KG (2007) The ripple effect: biofuels, food security, and the environment. Environment 49(9):30–43
Nepstad DC, Stickler CM, Soares-Filho B, Merry F (2008) Interactions among Amazon land use, forests and climate: prospects for a near-term forest tipping point. Philos T R Soc B 363: 1737–1746
Neushul P (1989) Seaweed for war: California's World War I kelp industry. Technol Cult 30: 561–583

Neushul P, Badash L (1998) Harvesting the Pacific: the blue revolution in China and the Philippines. Osiris 13:186–209
Neushul P, Wang Z (2000) Between the devil and the deep sea: C.K. Tseng, mariculture, and the politics of science in modern China. Isis 91:59–88
Ng HD, Lee JHS (2008) Comments on explosion problems for hydrogen safety. J Loss Prevent Proc 21:136–146
Ng TK, Busche RM, McDonald CC, Hardy RWF (1983) Production of feedstock chemicals. Science 219:733–740
Nguyen TLT, Gheewala SH, Garivait S (2008) Full chain energy analysis of fuel ethanol from cane molasses in Thailand. Appl Energ 85:722–734
Ni M, Leung DYC, Leung MKH (2007) A review on reforming bio-ethanol for hydrogen production. Int J Hydrogen Energ 32:3238–3247
Nicholls K, Campos S (2007) Are you driving on blood fuel? Ecologist 37(7):44–48
Numan MT, Bhosle NB (2006) $\alpha$-L-Arabinofuranosidases: the potential applications in biotechnology. J Ind Microbiol Biotechnol 33:247–260
Odada EO, Olago DO (2006) Challenges of an ecosystem approach to water monitoring and management of the African Great Lakes. Aquatic Ecosystem Health & Management 9:433–446
OECD (2008) Economic assessment of biofuel support policies. OECD directorate for trade and agriculture, Paris
OECD-FAO (2007) OECD-FAO Agricultural Outlook 2007–2016. OECD, Paris
Olofsson K, Bertilsson M, Lidén G (2008) A short review on SSF – an interesting process option for ethanol production from lignocellulosic feedstocks. Biotechnol Biofuels 1:7
Ooi Y, Bhatia S (2007) Aluminum-containing SBA-15 as cracking catalyst for the production of biofuel from waste used palm oil. Micropor Mesopor Mat 102:310–317
Oxfam (2007) Biofuelling poverty. Oxfam, Oxford
Özbay N, Oktar N, Tapan NA (2008) Esterification of free fatty acids in waste cooking oils (WCO): role of ion-exchange resins. Fuel 87:1789–1798
Pahkala K, Aalto M, Isolahti M, Poikola J, Jauhiainen L (2008) Large-scale energy grass farming for power plants – a case study from Ostrobothnia, Finland. Biomass Bioenerg 32:1009–1015
Palmarola-Adrados B, Choteborská P, Galbe M, Zacchi G (2005) Ethanol production from non-starch carbohydrates of wheat bran. Bioresour Technol 96:843–850
Park Y, Lee D, Kim D, Lee J, Lee K (2008) The heterogeneous catalyst system for the continuous conversion of free fatty acids in used vegetable oils for the production of biodiesel. Catal Today 131:238–243
Pigou AC (1920) The economics of welfare. MacMillan Company, London
Piringer G, Steinberg LJ (2006) Reevaluation of energy use in wheat production in the United States. J Ind Ecol 10:149–167
Prasad S, Singh A, Joshi HC (2007) Ethanol as an alternative fuel from agricultural, industrial and urban residues. Resour Conserv Recy 50:1–39
Ptasinski KJ, Hamelinck C, Kerkhof PJAM (2002) Exergy analysis of methanol from the sewage sludge process. Energ Convers Manage 43:1445–1457
Quadrelli R, Peterson S (2007) The energy–climate challenge: recent trends in $CO_2$ emissions from fuel combustion. Energ Policy 35:5938–5952
Qureshi N, Blaschek HP (2001) ABE production from corn: a recent economic evaluation. J Ind Microbiol Bitechnol 27:292–297
Qureshi N, Saha BC, Hector RE, Hughes SR, Cotta MA (2008a) Butanol production from wheat straw by simultaneous saccharification and fermentation using *Clostridium beijerinckii*. Biomass Bioenerg 32:168–175
Qureshi N, Ezeji TC, Ebener J, Dien BS, Cotta MA, Blaschek HP (2008b) Butanol production by *Clostridium beijerinckii*. Part I: use of acid and enzyme hydrolyzed corn fiber. Bioresour Technol 99:5915–5922
Qureshi N, Saha BC, Hector RE, Cotta MA (2008c) Removal of fermentation inhibitors from alkaline peroxide pretreated and enzymatically hydrolyzed wheat straw: production of butanol from hydrolysate using *Clostridium beijerinckii* in batch reactors. Biomass Bioenerg in press

Radich A (2004) Biodiesel performance, costs, and use. http://www.eia.doe.gov/oiaf/analysispaper/biodiesel/
Ramos MJ, Fernández CM, Casas A, Rodríguez L, Pérez A (2009) Influence of fatty acid composition of raw materials on biodiesel properties. Bioresour Technol 100:261–268
Ranganathan SV, Narasimhan SL, Muthukumar K (2008) An overview of enzymatic production of biodiesel. Bioresour Technol 99:3975–3981
Rantanen L, Linnaila R, Aakko P, Harju T (2005) NExBTL–Biodiesel fuel of the second generation. SAE International. http://www.nesteoil.com/default.asp?path=1,41,539,7516,7522
Rasi S, Veijanen A, Rintala J (2007) Trace compounds of biogas from different biogas production plants. Energy 32:1375–1380
Reed TB, Lerner RM (1973) Methanol: a versatile fuel for immediate use. Science 182:1299–1304
Reijnders L, Huijbregts MAJ (2005) Life cycle emissions of greenhouse gases associated with burning animal wastes in countries of the European Union. J Clean Prod 13:51–56
Reijnders L, Huijbregts MAJ (2007) Life cycle greenhouse gas emissions, fossil fuel demand and solar energy conversion efficiency in European bioethanol production for automotive purposes. J Clean Prod 15:1806–1812
Reijnders L, Huijbregts MAJ (2008) Palm oil and the emission of carbon-based greenhouse gases. J Clean Prod 16:477–482
Reinhardt G, Gärtner S, Patyk A, Rettenmaier N (2006) Ökobilanzen zu BTL: Eine ökologische Einschätzung. [Biomass to liquids: an environmental assessment]. Institut für Energie- und Umweltforschung Heidelberg GmbH, Heidelberg
Reinharz J (1985) Science in the service of politics: the case of Chaim Weizmann during the First World War. Engl Hist Rev 100:572–603
Renewable Fuels Agency (2008) The Gallagher review of the indirect effects of biofuels. Renewable Fuels Agency, St Leonards-on-Sea (East Sussex, UK)
Roberts MC (2008) E85 and fuel efficiency: an empirical analysis of 2007 EPA test data. Energ Policy 36:1233–1235
Rodríguez J, Lema JM, Kleerebezem R (2008) Energy-based models for environmental biotechnology. Trends Biotechnol 26:366–374
Ros J, Nagelhout D, Montfoort J (2009) New environmental policy for system innovation: Casus alternatives for fossil motor fuels. Appl Energ 86:243–250
Rosing MT, Bird DK, Sleep NH, Glassley W, Albarede F (2006) The rise of continents – an essay on the geologic consequences of photosynthesis. Palaeogeogr Palaeocl 232:99–113
Rotman D (2008) The price of biofuels. Technol Rev January/February:42–51
Rowlands WN, Masters A, Maschmeyer T (2008) The biorefinery – challenges, opportunities, and an Australian perspective. B Sci Technol Soc 28:149–158
Royal Society (2008) Sustainable biofuels: prospects and challenges. http://royalsociety.org
Rubin EM (2008) Genomics of cellulosic biofuels. Nature 454:841–845
Runge CF, Senauer B (2007) How biofuels could starve the poor. Foreign Affairs May/June
Rupprecht J, Hankamer B, Mussgnug JH, Ananyev G, Dismukes C, Kruse O (2006) Perspectives and advances of biological $H_2$ production in microorganisms. Appl Microbiol Biotechnol 72:442–449
Samaras C, Meisterling K (2008) Life cycle assessment of greenhouse gas emissions from plug-in hybrid vehicles: implications for policy. Environ Sci Technol 42:3170–3176
Sánchez OJ, Cardona CA (2008) Trends in biotechnological production of fuel ethanol from different feedstocks. Bioresour Technol 99:5270–5295
Sanderson K (2008) Flights of green fancy. Nature 453:264–265
Sassner P, Galbe M, Zacchi G (2008) Techno-economic evaluation of bioethanol production from three different lignocellulosic materials. Biomass Bioenerg 32:422–430
Sathaye J, Andrasko K (2007) Special issue on estimation of baselines and leakage in carbon mitigation forestry projects. Mitigation Adaptation Strategies Global Change 12:963–970
Saunders HD (2008) Virtual biofuels – a cheaper, better, faster alternative? Energ Policy 36: 1247–1250

Savage DF, Way J, Silver PA (2008) Defossiling fuel: how synthetic biology can transform biofuel production. ACS Chem Biol 3(1):13–16

Sawayama S, Minowa T, Yokoyama S (1999) Possibility of renewable energy production and $CO_2$ mitigation by thermochemical liquefaction of microalgae. Biomass Bioenerg 17:33–39

Scholz V, da Silva JN (2008) Prospects and risks of the use of castor oil as a fuel. Biomass Bioenerg 32:95–100

Schröder P, Herzig R, Bojinov A, Ruttens A, Nehnevajova E, Stamatiadis S, Memon A, Vassilev A, Caviezel M, Vangronsveld J (2008) Bioenergy to save the world. Producing novel energy plants for growth of abandoned land. Environ Sci Pollut Res Int 15:196–204

Scott A, Bryner M (2006) Alternative fuels: rolling out next-generation technologies. Chem Week 168:17–21

Scott E, Peter F, Sanders J (2007) Biomass in the manufacture of industrial products – the use of proteins and amino acids. Appl Microbiol Biotechnol 75:751–762

Scragg AH, Illman AM, Carden A, Shales SW (2002) Growth of microalgae with increased calorific values in a tubular reactor. Biomass Bioenerg 23:67–73

Semelsberger TA, Borup RL, Greene HL (2006) Dimethyl ether (DME) as an alternative fuel. J Power Sources 156:497–511

Shao H, Chu L (2008) Resource evaluation of typical energy plants and possible functional zone planning in China. Biomass Bioenerg 32:283–288

Shu Q, Yang B, Yuan H, Qing S, Zhu G (2007) Synthesis of biodiesel from soybean oil and methanol catalyzed by zeolite beta modified with $La^{3+}$. Catal Commun 8:2159–2165

Silver WL, Ostertag R, Lugo AE (2000) The potential for carbon sequestration through reforestation of abandoned tropical agricultural and pasture lands. Restor Ecol 8:394–407

Simonetti DA, Kunkes EL, Dumesic JA (2007) Gas-phase conversion of glycerol to synthesis gas over carbon-supported platinum and platinum–rhenium catalysts. J Catal 247:298–306

Sims REH, Hastings A, Schlamadinger B, Taylor G, Smith P (2006) Energy crops: current status and future prospects. Glob Change Biol 12:2054–2076

Sinclair TR, Muchow RC (1999) Radiation use efficiency. Adv Agron 65:215–265

Smeets E, Junginger M, Faaij A, Walter A, Dolzan P, Turkenburg W (2008) The sustainability of Brazilian ethanol – an assessment of the possibilities of certified production. Biomass Bioenerg 32:781–813

Solomon BD, Barnes JR, Halvorsen KE (2007) Grain and cellulosic ethanol: history, economics, and energy policy. Biomass Bioenerg 31:416–425

Song C, Zhou Y, Huang R, Wang Y, Huang Q, Lu G, Liu K (2007) Influence of ethanol-diesel blended fuels on diesel exhaust emissions and mutagenic and genotoxic activities of particulate extracts. J Hazard Mater 149:355–363

Sørensen A, Teller PJ, Hilstrøm T, Ahring BK (2008) Hydrolysis of *Miscanthus* for bioethanol production using dilute acid presoaking combined with wet explosion pre-treatment and enzymatic treatment. Bioresour Technol 99:6602–6607

Stams AJM, de Bok FAM, Plugge CM, van Eekert MHA, Dolfing J, Schraa G (2006) Exocellular electron transfer in anaerobic microbial communities. Environ Microbiol 8:371–382

Steiner C, Teixeira WG, Lehmann J, Nehls T, de Macêdo JLV, Blum WEH, Zech W (2007) Long term effects of manure, charcoal and mineral fertilization on crop production and fertility on a highly weathered Central Amazonian upland soil. Plant Soil 291:275–290

Subramani V, Gangwal SK (2008) A review of recent literature to search for an efficient catalytic process for the conversion of syngas to ethanol. Energ Fuels 22:814–839

Sujatha M, Reddy TP, Mahasi MJ (2008) Role of biotechnological interventions in the improvement of castor (*Ricinus communis* L.) and *Jatropha curcas* L. Biotechnol Adv 26:424–435

Szklo A, Schaeffer R, Delgado F (2007) Can one say ethanol is a real threat to gasoline? Energ Policy 35:5411–5421

Takeshita T, Yamaji K (2008) Important roles of Fischer-Tropsch synfuels in the global energy future. Energ Policy 36:2773–2784

Tamunaidu P, Bhatia S (2007) Catalytic cracking of palm oil for the production of biofuels: optimization studies. Bioresour Technol 98:3593–3601

Tang H, Salley SO, Ng KYS (2008) Fuel properties and precipitate formation at low temperature in soy-, cottonseed-, and poultry fat-based biodiesel blends. Fuel 87:3006–3017

Taylor LE II, Dai Z, Decker SR, Brunecky R, Adney WS, Ding S, Himmel ME (2008) Heterologous expression of glycosyl hydrolases *in planta*: a new departure for biofuels. Trends Biotechnol 26:413–424

Themelis NJ, Ulloa PA (2007) Methane generation in landfills. Renew Energ 32:1243–1257

Tilman D, Hill J, Lehman C (2006) Carbon-negative biofuels from low-input high-diversity grassland biomass. Science 314:1598–1600

Tollefson J (2008) Energy: not your father's biofuels. Nature 451:880–883

Troell M, Robertson-Andersson D, Anderson RJ, Bolton JJ, Maneveldt G, Halling C, Probyn T (2006) Abalone farming in South Africa: an overview with perspectives on kelp resources, abalone feed, potential for on-farm seaweed production and socio-economic importance. Aquaculture 257:266–281

Tsuchida T, Yoshioka T, Sakuma S, Takeguchi T, Ueda W (2008) Synthesis of biogasoline from ethanol over hydroxyapatite catalyst. Ind Eng Chem Res 47:1443–1452

Tsutsui WM (2003) Landscapes in the dark valley: toward an environmental history of wartime Japan. Environ Hist 8:294–311

Turner P, Mamo G, Nordberg Karlsson E (2007) Potential and utilization of thermophiles and thermostable enzymes in biorefining. Microb Cell Fact 6:9

Tyner WE (2007) Policy alternatives for the future biofuels industry. J Agric Food Ind Organ 5(2):2. http://www.bepress.com/jafio/vol5/iss2/art2/

Tyner WE (2008) The US ethanol and biofuels boom: its origins, current status, and future prospects. BioScience 58:646–653

Ugwu CU, Aoyagi H, Uchiyama H (2008) Photobioreactors for mass cultivation of algae. Bioresour Technol 99:4021–4028

Valliyappan T, Bakhshi NN, Dalai AK (2008) Pyrolysis of glycerol for the production of hydrogen or syn gas. Bioresour Technol 99:4476–4483

van Dam J, Junginger M, Faaij A, Jürgens I, Best G, Fritsche U (2008) Overview of recent developments in sustainable biomass certification. Biomass Bioenerg 32:749–780

Van Ginkel SW, Oh S, Logan BE (2005) Biohydrogen gas production from food processing and domestic wastewaters. Int J Hydrogen Energ 30:1535–1542

Vasudevan PT, Briggs M (2008) Biodiesel production – current state of the art and challenges. J Ind Microbiol Biotechnol 35:421–430

Von Braun J (2008) Food prices, biofuels and climate change. International Food Policy Research Institute, Washington DC

Wackett LP (2008) Biomass to fuels via microbial transformations. Curr Opin Chem Biol 12: 187–193

Wang J, Shuai S, Chen H (2007) Application and development of biomass fuels for transportation in China. Tsinghua Sci Technol 12:223–230

Wang C, Pan J, Li J, Yang Z (2008) Comparative studies of products produced from four different biomass samples via deoxy-liquefaction. Bioresour Technol 99:2778–2786

Wang D, Bean S, McLaren J, Seib P, Madl R, Tuinstra M, Shi Y, Lenz M, Wu X, Zhao R (2008a) Grain sorghum is a viable feedstock for ethanol production. J Ind Microbiol Biotechnol 35: 313–330

Wang L, Weller CL, Jones DD, Hanna MA (2008b) Contemporary issues in thermal gasification of biomass and its application to electricity and fuel production. Biomass Bioenerg 32:573–581

Wang H, Brown SL, Magesan GN, Slade AH, Quintern M, Clinton PW, Payn TW (2008c) Technological options for the management of biosolids. Environ Sci Pollut Res 15:308–317

Wardle DA (2003) Global sale of green air travel supported using biodiesel. Renew Sust Energ Rev 7:1–64

Wardle DA, Nilsson M, Zackrisson O (2008) Fire-derived charcoal causes loss of forest humus. Science 320:629

Warnick TA, Methé BA, Leschine SB (2002) Clostridium phytofermentans sp. nov., a cellulolytic mesophile from forest soil. Int J Syst Evol Microbiol 52:1155–1160

Weng J, Li X, Bonawitz ND, Chapple C (2008) Emerging strategies of lignin engineering and degradation for cellulosic biofuel production. Curr Opin Biotechnol 19:166–172

Westermann P, Jørgensen B, Lange L, Ahring BK, Christensen CH (2007) Maximizing renewable hydrogen production from biomass in a bio/catalytic refinery. Int J Hydrogen Energ 32: 4135–4141

Westhoff P (2008) Farm commodity prices: why the boom and what happens now? Choices 23(2):6–10

Wheals AE, Basso LC, Alves DMG, Amorim HV (1999) Fuel ethanol after 25 years. Trends Biotechnol 17:482–487

Wiesenthal T, Leduc G, Christidis P, Schade B, Pelkmans L, Govaerts L, Georgopoulos P (2008) Biofuel support policies in Europe: lessons learnt for the long way ahead. Renew Sust Energ Rev in press

Wijffels RR (2008) Potential of sponges and microalgae for marine biotechnology. Trends Biotechnol 26:26–31

Wikfors GH, Ohno M (2001) Impact of algal research in aquaculture. J Phycol 37:968–974

Wilcox HA (1982) The ocean as a supplier of food and energy. Experientia 38:31–35

Wilhelm SW, Suttle CA (1999) Viruses and nutrient cycles in the sea. BioScience 49:781–788

Willke T, Vorlop K (2004) Industrial bioconversion of renewable resources as an alternative to conventional chemistry. Appl Microbiol Biotechnol 66:131–142

Winebrake JJ, Corbett JJ, Meyer PE (2007) Energy use and emissions from marine vessels: a total fuel life cycle approach. J Air Waste Manag Assoc 57:102–110

Wongtanet J, Sang B, Lee S, Pak D (2007) Biohydrogen production by fermentative process in continuous stirred-tank reactor. Int J Green Energy 4:385–395

Würster R, Zittel W (2007) Hydrogen infrastructure build-up for automotive applications. Mitigation Adaptation Strategies Global Change 12:367–386

Wyman CE (2007) What is (and is not) vital to advancing cellulosic ethanol. Trends Biotechnol 25:153–157

Yang YF, Feng CP, Inamori Y, Maekawa T (2004) Analysis of energy conversion characteristics in liquefaction of algae. Resour Conserv Recy 43:21–33

Yazdani SS, Gonzalez R (2007) Anaerobic fermentation of glycerol: a path to economic viability for the biofuels industry. Curr Opin Biotechnol 18:213–219

You Y, Shie J, Chang C, Huang S, Pai C, Yu Y, Chang CH (2008) Economic cost analysis of biodiesel production: case in soybean oil. Energy Fuels 22:182–189

Yumrutaş R, Alma MH, Özcan H, Kaşka Ö (2008) Investigation of purified sulphate turpentine on engine performance and exhaust emission. Fuel 87:252–259

Zaldivar J, Nielsen J, Olsson L (2001) Fuel ethanol production from lignocellulose: a challenge for metabolic engineering and process integration. Appl Microbiol Biotechnol 56:17–34

Zanichelli D, Carloni F, Hasanaji E, D'Andrea N, Filippini A, Setti L (2007) Production of ethanol by an integrated valorization of olive oil byproducts: the role of phenolic inhibition. Environ Sci Pollut Res Int 14:5–6

Zhou X, Xiao B, Ochieng RM, Yang J (2008) Utilization of carbon-negative biofuels from low-input high-diversity grassland biomass for energy in China. Renew Sust Energ Rev in press

Zverlov VV, Berezina O, Velikodvorskaya GA, Schwarz WH (2006) Bacterial acetone and butanol production by industrial fermentation in the Soviet Union: use of hydrolyzed agricultural waste for biorefinery. Appl Microbiol Biotechnol 71:587–597

# Chapter 2
# Energy Balance: Cumulative Fossil Fuel Demand and Solar Energy Conversion Efficiency of Transport Biofuels

## 2.1 Introduction

The adjective 'sustainable' is frequently used regarding biofuels (e.g. Abrahamson et al. 1998; Krotscheck et al. 2000; Buckley and Schwarz 2003; Bhattacharya et al. 2003; Goldemberg and Teixeira Coelho 2004; Butterworth 2006; Demirbaş 2007; Robèrt et al. 2007; Goldemberg et al. 2008; Karp and Shield 2008; Royal Society 2008). Also, biofuels are a regular subject in scientific journals dealing with renewable or sustainable energy. The apparent rationale of using 'sustainable' and 'renewable' in the context of biofuels is the following: biomass may be argued to temporarily store solar energy, based on photosynthesis (see Chap. 1). In doing so, carbon is sequestered, and on burning transport biofuel, it is de-sequestered. In the meantime, photosynthesis proceeds, generating new feedstocks for biofuels. As solar irradiation and photosynthesis are expected to last for many millions of year, doing so would seem sustainable and transport biofuels renewable. However, this is not the 'whole story'. Energy inputs in the world economy are currently, as pointed out in Chap. 1, overwhelmingly fossil fuels, and the use of fossil fuels extends to the production and distribution of transport biofuels. This is at variance with renewability and sustainability, as fossil fuels are non-renewables, and their use cannot be sustained indefinitely at the present level. The cumulative life cycle fossil fuel demand of biofuels will be discussed in Sect. 2.3.

For converting solar irradiation into transport kilometres, there are a variety of technologies available with widely varying efficiencies. Such efficiencies matter: they are major determinants of spatial requirements of energy supply. These spatial requirements, in turn, are important determinants of competition of energy supply with food production and habitats for living nature. Because this competition is an important matter in the current transport biofuel debate and will return later in this book, this chapter will deal with the solar conversion efficiency of transport biofuels (Sects. 2.4 and 2.5). Other methods for solar energy conversion do not involve organisms but rely on physical conversion technologies. Photovoltaic cells generating electricity are examples thereof, for which the solar conversion efficiency will be discussed in Sect. 2.6.

L. Reijnders, M.A.J. Huibregts, *Biofuls for Road Transport*

Biofuels and the output from photovoltaic cells can be used to perform work or to deliver energy services. Work (a thermodynamic concept) or energy services (an economic concept) include, for instance, car kilometres. The performance of work often includes the use of intermediaries (e.g. power plants, batteries or motors). The energy efficiencies of such intermediaries will be discussed in Sect. 2.6. In Sect. 2.6, we will also consider the overall efficiency of a variety of methods to convert solar energy to car kilometres, giving a 'seed-to-wheel' perspective.

As pointed out in Chap. 1, the production of biofuels is often accompanied by by-products or co-products. For instance, in making biodiesel from rapeseed, both glycerol and an ingredient of animal feed (rapeseed cake) are produced. Before we go into the calculations of cumulative (fossil) energy demand, it should be decided how much of that demand is allotted to biodiesel and how much to glycerol and rapeseed cake. This is called 'allocation' and will be discussed in Sect. 2.2.

## 2.2 Allocation

There are three major ways to allocate. The first is based on prices, the second on physical categories, such as weight or energy, and the third on subtracting avoided processes (also called substitution). We will look at these in turn. The first way to allocate is on the basis of price (market values). The idea behind this type of allocation is that prices drive production (Weidema 1993). This method is, however, not without problems. Firstly, market prices are not constants. So, if, for example, ethanol prices go up, whereas the prices of other outputs do not, the emissions and cumulative fossil energy demand allocated to this transport fuel increase. The same happens when by-products go down in price, but the transport biofuel price remains constant (or increases). A good example of the latter is the tenfold price decrease of glycerol between 2004 and 2006 (Yazdani and Gonzalez 2007).

A second problem is that currently, much transport biofuel production is not driven by market value but by market value plus subsidy. This leads to the question of whether, for instance, in the case of ethanol production from cornstarch, allocation should be on the basis of the market value of cornstarch or on the basis of the subsidized value. Another problem arises when wastes are considered. These may well have negative prices (being a cost to the producer). For instance, the producer of the waste may have to pay a price for the incineration or treatment of his waste. If so, allocation on the basis of price may mean that the waste, because of its negative price, is apparently associated with a negative cumulative energy demand (Reijnders and Huijbregts 2005). Usually, this has been felt unsatisfactory by proponents of allocation based on prices, and this often leads to the decision the give a zero price to wastes. However, this seems inconsistent. An implication of a zero price is that the life cycle leading to the generation of wastes has no impact on the environmental evaluation of such biofuels. The problem may also arise as to whether something is a waste or a by-product. An example thereof is sawdust. This may be used for firing industrial installations or power plants, and then may be categorized as by-product (with a positive monetary value), but sawdust may also be left in the woods and may then be categorized as a waste (with zero monetary value). Decisions regarding such categorizations may be far from easy and may have a substantial impact on the greenhouse gas emissions calculated.

Alternatively, one may allocate on the basis of physical categories such as 'energy content' (heating value) or weight. For instance, the European Union in its 2008 draft Renewables Directive has proposed to allocate on the basis of energy (Eickhout et al. 2008). This type of allocation has the advantage of stable outcomes, unaffected by movements of prices. However, there are curious consequences, too. For instance, in this allocation system, there is an obvious way to improve the environmental performance of a transport biofuel, and that is to produce more waste. To evade this problem, there is a tendency to restrict allocation to product outputs. Matters related to quality may also emerge. If one, for instance, allocates to the outputs of electricity and low temperature heat on the basis of 'energy content', one may be criticized for neglecting the quality of these outputs and be advised to use exergy instead of energy. Thus, allocation on the basis of physical categories may encounter criticism if the physical property chosen is at variance with the perceived value of the co-products.

Another way to deal with a multi-output process is to 'correct the system'. In the case of biofuels, one may consider biofuel to be the only output and correct for the other outputs by subtracting 'avoided processes' which such outputs can substitute (Ekvall and Finnveden 2001). This approach has also been called substitution. For instance, in the case of ethanol production from corn or wheat, it has been argued that by-products such as dried distillers grains (DDG) or dried distillers grains with solubles (DDGS) may be a substitute of soybean meal in cattle feed (Kim and Dale 2005). Thus, producing DDG(S) may be valued on the basis of the avoided process of producing soybean meal. However, soybean meal and DDG(S) are not identical. This then raises the question of the basis for conversion: should it be on the basis of price, or protein content, or metabolizable joules (energy)? Moreover, DDG(S) is not a straightforward substitute of soybean meal, as its composition is relatively variable, and its consumption by animals may be linked to increased mycotoxicosis risk and increased intakes of mycotoxins (Taylor-Pickard 2008). This has led to a more limited recommended use of DDG(S) in animal feed than in the case of soybean meal (Taylor-Pickard 2008). Then there is the matter of applications other than animal feed. For instance, soybean meal may also be used to generate vegetarian alternatives to meat, and DDG(S) may be used to produce protease and peptones (Romero et al. 2007), methane (Murphy and Power 2008) or ethanol. Such alternative applications may have environmental impacts that are very different from the use as an ingredient of animal feed. Suppose, finally, that DDG(S) is indeed valued on the basis of avoiding soybean meal; the problem is that soybean meal is a co-product, just as DDG(S) is. This may be argued to imply that substitution in this case means plugging one hole with another.

So, each way to allocate has its weak points, and there is no agreement on the best way to allocate. In this book, we will not make a choice in favour of a specific way to allocate but rather will explicitly indicate what type of allocation has been used in arriving at specific results.

## 2.3 Cumulative Fossil Fuel Demand

### *2.3.1 Transport Biofuels from Terrestrial Plants*

Most studies regarding cumulative fossil energy demand have been done for transport biofuels from terrestrial plants, and most agree that the seed-to-wheel cumulative demand for fossil fuels associated with transport biofuels from terrestrial plants is lower than the well-to-wheel demand of fossil transport fuels. However, Patzek and Pimentel (Pimentel 2003; Patzek 2004; Patzek and Pimentel 2005; Patzek 2006) have presented calculations for cornstarch-derived ethanol, soybean- and sunflower-derived biodiesel and lignocellulosic ethanol that suggest a higher cumulative demand for fossil fuels. The difference between these studies of Patzek and Pimentel and other studies is partly caused by difference in allocation, partly by higher estimates of fossil fuel input in agriculture and industrial processing, and partly by factoring in the energy demand of the infrastructure needed for transport biofuel production (factories, vehicles, etc.) into the calculations. However, along with assumptions that are more favourable to transport biofuels, there seems no denying that in western industrialized countries, the cumulative fossil energy demand for transport biofuels made from starch, sugar and edible oils may be quite high when allocation is on the basis of price. For ethanol from US corn or European wheat or rye, it would seem unlikely that, when allocated on this basis, the 'seed-to-wheel' cumulative fossil energy demand would be much lower than 80% of the corresponding demand for petrol (Hammerschlag 2006; Hill et al. 2006; von Blottnitz and Curran 2007; Reijnders and Huijbregts 2007; Zah et al. 2007). In the case of biodiesel from rapeseed and soybean, qualitatively good estimates usually suggest that, when allocated on the basis of price, the cumulative energy demand may well be in the order of 60–80% of the corresponding demand for diesel (Hill et al. 2006; Zah et al. 2007).

Cumulative fossil energy demand for transport biofuels may be considerably lower when biofuels based on high-yielding crops from developing counties, such as oil palm and sugar cane, are considered, especially when lignocellulosic biomass is used for powering processing facilities (von Blottnitz and Curran 2007; Reijnders and Huijbregts 2008a). When the latter applies, for instance, cumulative fossil fuel inputs in ethanol from sugar cane may become energetically less than 10% of the ethanol output (Macedo et al. 2008). Also, much lower cumulative fossil fuel demands have been estimated for transport biofuels from lignocellulosic biomass such as wood or switchgrass when processing is also powered by lignocellulosic biomass (von Blottnitz and Curran 2007). When allocation is based on the energy content or weight of outputs, cumulative fossil energy demand allocated to transport biofuels will tend to be lower than in the case of allocation based on price. Note that cumulative mineral oil demand is often lower than cumulative fossil fuel demand, because natural gas and coal can be significant contributors of energy to the transport biofuel life cycle (Hammerschlag 2006; Kim and Dale 2008). For instance, coal is often an important contributor to electricity supply, which is sometimes used by mills producing ethanol (Kim and Dale 2008). Natural gas is important in production of fixed nitrogen to be used in agriculture (Hammerschlag 2006).

### 2.3.2 Transport Biofuels from Wastes

Zah et al. (2007) have studied cumulative energy demand associated with methane production from a variety of wastes using allocation on the basis of price and a zero value for the waste itself. Thus, the calculation of energy demand and emissions linked to methane production from wastes was restricted to the waste-to-wheel stages of the life cycle. Comparison was with natural gas. The wastes considered were: sewage sludge, 'biowaste', manure and manure plus co-substrates. Cumulative fossil energy demand for methane from these wastes was typically in the order of approximately 45% of the fossil reference. The outcomes of the study of Zah et al. (2007) are more favourable to transport biofuels made from wastes than to transport biofuels made from food crops. Zwart et al. (2006) made a more detailed study of the conversion of manure from cattle and swine into biogas (methane) in the Netherlands and concluded that the fossil fuel input energetically roughly equalled the biogas output. One should keep in mind that these outcomes are based on the assumption that life cycle impacts up to the waste can be neglected. When wastes change into secondary resources, fetching a price, or when the seed-to-wheel allocation is based on mass or energy, this would raise cumulative fossil energy demand of transport biofuels made from residues (cf. Reijnders and Huijbregts 2005).

### 2.3.3 Transport Biofuels from Aquatic Biomass

Fossil fuel inputs in producing microalgae tend to be high. When microalgae are grown in bioreactors, outputs are unlikely to energetically outperform inputs (Wijffels 2008; Reijnders 2008). A claim has been made for ultrahigh bioproductivity from algae in thin channel ultradense culture bioreactors indirectly irradiated by the sun (Gordon and Polle 2007). The cultures are irradiated with pulsed light emitting diodes, powered by photovoltaic cells. The efficiency of converting solar radiation into biomass is probably below 0.2%, and the corresponding energetic yield is likely to be exceeded by fossil fuel inputs (Wijffels 2008).

As to producing microalgal biofuels in open ponds, it is a remarkable aspect of several recent publications strongly advocating algal transport biofuels (e.g. Chisti 2007; Huntley and Redalje 2007; Chisti 2008a; Dismukes et al. 2008) that inputs of fossil fuels are not addressed. Two less recent studies are available that looked at energy inputs and outputs in open pond cultures of microalgae. They did not take account of all inputs, though. For instance, fossil fuel input into the handling and clean-up of discharges from ponds (which will probably be necessary in view of the extreme pH and/or salt concentrations and high nutrient levels in algal ponds) was considered by neither of the studies. Sawayama et al. (1999) studied operational life cycle energy inputs in growing and processing *Dunaliella tertiolecta* to supply bio-oil. Processing was by thermal liquefaction (also Yang et al. 2004). Operational energy inputs (fossil fuels) exceeded energetic output by 56% when microalgal yield was 15 Mg $ha^{-1}$ $year^{-1}$. Hirano et al. (1998) studied *Spirulina* production and processing to supply methanol (via synthesis gas). Here the assumed yield was approximately 110 Mg $ha^{-1}$ $year^{-1}$. Both fossil fuel inputs in infrastructure and operation were considered. The energetic output exceeded the life cycle fossil fuel input by 10%. At more realistic estimates of *Spirulina* yield, which are in the order of 10–30 Mg $ha^{-1}$ $year^{-1}$ (Vonshak and Richmond 1988; Jiménez et al. 2003), fossil fuel inputs would have exceeded energetic outputs. Chisti (2008b) has argued that the energetic inputs used in the studies of Hirano et al. (1998) and Sawayama et al. (1999) are 'grossly overestimated'. However, even at Chisti's (2008b) estimate, the fossil fuel input energetically would equal an output of approximately 30 Mg dry weight algal biomass $ha^{-1}$ $year^{-1}$, which is at the upper end of the range for the commercial production of *Spirulina* (Jiménez et al. 2003).

Though experimentally, yields have been demonstrated that may energetically exceed fossil fuel inputs (Hirano et al. 1998; Chisti 2008b), it is far from certain that such yields can be achieved in actual commercial practice. Large differences between experimental yields and average commercial yields are also common in the production of terrestrial crops, as will be explained in Sect. 2.4.1.

A 'high yield' has furthermore been claimed for oil from *Haematococcus pluvialis* produced by a combination of a closed bioreactor and 1.3 days in a pond (Huntley and Redalje 2007). This yield probably corresponds with a photosynthetic efficiency in producing biomass of just over 1% and a photosynthetic efficiency in producing algal oil of roughly 0.6% (Vasudevan and Briggs 2008). No data have been published about the cumulative energetic inputs in this type of culture, but from the above, it would seem unlikely that the energetic value of algal oil would much exceed the cumulative energy input into the infrastructural and operational inputs.

Studies regarding algal production of $H_2$ suggest that the cumulative energy demand for algal $H_2$ production is probably of the same order of magnitude as the energetic output, when the solar energy conversion efficiency does not exceed 1% (Burgess and Fernández-Velasco 2007).

On the other hand, it may be that the yield of microalgae grown in water saturated by $CO_2$ from power stations may exceed fossil fuel inputs when there is no allocation of the fossil fuel input into electricity production to these algae. However, whether this application will actually become operational is unclear, as algal performance has so far been disappointing, and sequestration of $CO_2$ in abandoned gas and oil fields and aquifers has a higher efficiency (Benemann et al. 2003; Vunjak-Novakovic et al. 2005; Odeh and Cockerill 2008).

The emergence of some saltwater and freshwater macroalgae and macrophytes as pests offers scope for their conversion into transport biofuels. Only for one of the macrophytes (water hyacinth) are data available about the overall energy efficiency of conversion into ethanol. These data suggest a negative energy balance (Gunnarsson and Petersen 2007).

## 2.4 Conversion of Solar Energy into Biomass

The intercept by the Earth of solar energy exceeds the present input of fossil and uranium fuels into the world economy by a factor of about 10,000 (Lewis and Nocera 2006). The average daily solar irradiation varies, dependent on latitude, climate and season. When on the equator, maximum irradiation is on a horizontal plane, but away from the equator, for the maximum intercept of solar radiation by a fixed plane, the plane should have an angle corresponding to latitude (e.g. Çelik 2006). Average daily solar irradiation (measured on a horizontal surface) that may support feedstock for biofuel production varies roughly between 7 and 25 $\mathrm{MJ\,m^{-2}}$. The daily worldwide average irradiation is about 15.5 $\mathrm{MJ\,m^{-2}}$, or 180 $\mathrm{W\,m^{-2}}$. Differences between days can be large. For instance, in Amsterdam (52°21′ N), the average daily irradiation is approximately 3 $\mathrm{MJ\,m^{-2}}$ in January and 17 $\mathrm{MJ\,m^{-2}}$ in July (Akkerman et al. 2002). The greatest annual input of solar radiation tends to occur in subtropical regions at latitudes between 20 and 30° and little cloud cover. Humid tropical regions have somewhat lower irradiation (Sinclair and Muchow 1999). When going poleward from a latitude of about 30°, solar irradiation tends to decrease. As for major areas for current biofuel production, in Brazil, where sugar cane ethanol is produced, daily solar irradiation is on average about 220 $\mathrm{W\,m^{-2}}$ (approximately 19 $\mathrm{MJ\,day^{-1}\,m^{-2}}$ or $694 \times 10^2$ $\mathrm{GJ\,year^{-1}\,ha^{-1}}$). In the US, average daily irradiation varies between 12 and 22 $\mathrm{MJ\,m^{-2}}$, whereas in the US Midwest, where there is large-scale corn ethanol production, solar irradiation is about 170 $\mathrm{W\,m^{-2}}$ (approximately 14.7 $\mathrm{MJ\,day^{-1}\,m^{-2}}$ or $536 \times 10^2$ $\mathrm{GJ\,year^{-1}\,ha^{-1}}$) (Kheshgi et al. 2000; Vasudevan and Briggs 2008).

In establishing the overall conversion efficiency of technologies for the conversion of solar energy, there should be a correction for the cumulative energy demand associated with the biofuel life cycle and the life cycle of physical conversion technologies (Reijnders and Huijbregts 2007). For instance, if the lower heating value of fossil fuel inputs amounts to 20% of the lower heating value of a biofuel, the solar energy conversion efficiency will be corrected by this percentage. The result thereof is the overall energy efficiency of the biofuel. This is summarized in the following equation:

$$SCE_x = \frac{Y_x \cdot E_x \cdot FE_x}{E_{\mathrm{solar}}} \cdot 100$$

where $SCE_x$ is the solar energy conversion efficiency of biomass or biofuel type $x$ (%), $Y_x$ is the yield of biomass type $x$ (kg/ha/year), $E_x$ the energy content of biomass or biofuel type $x$ (MJ/kg), $FE_x$ the correction factor for fossil fuel input in the life cycle of biomass or biofuel type $x$ (MJ/MJ), and $E_{solar}$ is the yearly solar irradiation (MJ/ha/year). $SCE_x$ is a measure that can help in estimating the ability of biofuels to displace fossil fuels.

As pointed out in Chap. 1, conversion of solar energy into biomass occurs by photosynthesis. Harvestable biomass that can be used for energy generation (yield) depends on a number of factors. At the present atmospheric $CO_2$ concentration for $C_4$ terrestrial plants, the maximum conversion efficiency is estimated at 5.5–6.7% and for $C_3$ plants at 3.3–4.6% (Hall 1982; El Bassam 1998; Heaton et al. 2008b). A 6.7% solar energy conversion efficiency would correspond with a dry biomass yield of approximately 250 Mg ha$^{-1}$ year$^{-1}$ at 40° latitude (El Bassam 1998). Actual yields are much lower than theoretical yields, because there are factors – such as in the case of terrestrial plants, the absence of a full canopy, shading, photosaturation and limited availability of nutrients and water – which in practice reduce the efficiency. Due to such factors, the theoretical differences in conversion efficiency between, for instance, $C_3$ and $C_4$ plants may not materialize in real life differences in conversion efficiency. For instance, sorghum is a $C_4$ plant that tends to be roughly as efficient as the $C_3$ cereals. And $C_3$ plants such as sugar beet and oil palm are in practice often more efficient in converting solar radiation into biomass than the $C_4$ plant *Miscanthus*.

### *2.4.1 Terrestrial Plants*

Terrestrial plants vary widely in their yearly yields per hectare. Yields are dependent on insolation, temperature, the presence of nutrients and water and the nature of plants (Coombs et al. 1987). In natural ecosystems on average, the efficiency of photosynthesis in converting solar energy into plant material is usually in the order of 0.1–0.3% (Mezhunts and Givens 2004; Rosing et al. 2006). In the case of cultivated plants, higher conversion efficiencies are achievable. The highest yields are usually achieved in experiments under 'excellent' conditions that are highly conducive to plant growth. In large-scale commercial cultivation, yields are much lower. In the following, we will use data from large-scale cultivation, as this should be the basis for substantial feedstock production. As there is a tendency of gradual yield increases over time, such data may be biased in favour of crops that have a long tradition of large-scale cultivation. After a similar history of cultivation, the yields of relatively new crops that may serve as biofuel feedstocks such as *Miscanthus* and switchgrass may well be substantially higher than those that will be presented here.

In practice, the $C_4$ plant sugar cane is relatively efficient in converting solar energy into biomass (Sinclair and Muchow 1999). In subtropical areas, sugar cane may annually yield about 80 Mg per hectare of harvestable biomass (dry weight) when the conditions are excellent (Bastiaanssen and Ali 2003; Braunack et al. 2006). Average sugar cane yields during the mid 1990s in Brazil were about 36.8 tons of biomass (dry weight) ha$^{-1}$ year$^{-1}$ (Kheshgi et al. 2000). Under excellent conditions, another $C_4$ plant, *Miscanthus*, may yield annually up to about 30–60 Mg of dry weight harvestable biomass per hectare (Long et al. 2006; Heaton et al. 2008a), but more commonly, yields are in the range of 10–13 Mg aboveground dry weight biomass $ha^{-1}$ $year^{-1}$ (Lemus and Lal 2005; Christian et al. 2008).

Oil palms in Southeast Asia yield about 20 Mg $year^{-1}$ $ha^{-1}$ as fresh fruit bunches (dry weight) (Reijnders and Huijbregts 2008a). For sugar beets, good yearly dry weight yields of biomass from large-scale commercial cultivation are also in the order of 20 Mg $ha^{-1}$ (Şahin et al. 2004; Tzilivakis et al. 2005). For eucalyptus, yearly biomass yields per hectare tend to be in the order of 10–20 tons (Sims et al; 1999; van den Broek et al. 2001). Yearly dry biomass yields of large-scale cultivation under good conditions for switchgrass are in the order of 10–15 Mg $ha^{-1}$, for willow 9 Mg $ha^{-1}$, and for poplar 11 Mg $ha^{-1}$ (Lemus and Lal 2005; Heaton et al. 2008a). Total yearly (dry weight) aboveground biomass accumulation per hectare in the USA is in the order of 17–18 Mg for corn (Heaton et al. 2008a), and under good conditions, 10–11 Mg for wheat (world average is 5.5 Mg; Wright et al. 2001), in the order of 9 Mg for peas and 4–5 Mg for canola (Lemus and Lal 2005; Malhi et al. 2006). High yields of photosynthesis in practice usually depend on substantial inputs of synthetic nutrients derived from non-renewable natural resources (Samson et al. 2005). Sustainable yields that can be achieved when only recycling nutrients that are present in biomass tend to be much lower as will be discussed in Chap. 3 (also Pimentel et al. 2002; Reijnders 2006). Table 2.1 shows the overall energy conversion efficiency (taking account of inputs of fossil fuels) for a variety of crops with relatively good yields.

The overall solar energy conversion efficiencies in Table 2.1 are below 1% and range roughly between 0.15% (for rapeseed/canola) and 0.9% (for sugar cane). For comparison, a percentage is added for sustainably grown wood in Western Russia (Nabuurs and Lioubimov 2000). In this case, the conversion efficiency is about 0.05%.

There have been efforts to improve the solar-energy-to-biomass conversion by transgenic approaches. These have focused on increasing the net carboxylation efficiency of 1,5-biphosphate carboxylase and the introduction of enzymes characteristic for $C_4$ plants in $C_3$ plants (Heaton et al. 2008; Raines 2006). So far, such efforts have not led to a substantial improvement in the conversion of solar energy to biomass (Raines 2006).

### *2.4.2 Terrestrial Biofuels*

For some applications, biomass as it is produced in solar energy conversion may be used as such. This applies, for instance, to the generation of electricity, which in turn may be used for electrical traction. However, diesel or Otto motors or fuel cells need the use of specific biochemicals (transport biofuels) such as specific alcohols and acylesters, as discussed in Chap. 1. This has an impact on the efficiency of solar conversion. Only a part of the biomass originating in solar energy conversion can be turned into such chemicals. It may be that part of the biomass that cannot be con-

**Table 2.1** Solar energy to biomass conversion efficiencies, with correction for fossil fuel inputs

| Insolation (MJ/day $m^2$) | Crop under good conditions (unless otherwise indicated) | Yield of biomass $ha^{-1}$ (Mg dry weight/year); aboveground except for sugar beet | Energy content biomass (lower heating value in MJ/kg dry weight) | Correction factor for fossil fuel input (MJ in crop – MJ fossil fuel input/MJ in crop) | Solar energy conversion efficiency (%) |
|---|---|---|---|---|---|
| 19 | Sugar cane (average) | 36.8 (Kheshgi et al. 2000) | 17.5 | 0.97 (Dias de Oliveira et al. 2005) | 0.9 |
| 19 | Oil palm | 20 (fruit bunches) | 31.7 | 0.95 (Reijnders and Huijbregts 2003) | 0.87 |
| 19 | *Eucalyptus* | 10–20 | 19 | 0.9 (estimate) | 0.25–0.50 |
| 14 | Wheat | 10–11 | 17.5 | 0.8 (von Blottnitz and Curran 2007) | 0.27–0.30 |
| 14 | Switchgrass | 10–15 | 17.5 | 0.95 (estimate) | 0.32–0.48 |
| 14 | Sugar beet | 20 | 17 | 0.9 (von Blottnitz and Curran 2007) | 0.62 |
| 14 | Corn | 17–18 | 17.5 | 0.8 (von Blottnitz and Curran 2007) | 0.46–0.49 |
| 14 | Rapeseed/Canola | 4–5 | 21.8 | 0.9 (Zah et al. 2007) | 0.15 |
| 14 | *Miscanthus* | 10–13 | 17.5 | 0.98 (Lewandowski and Schmidt 2006) | 0.34–0.44 |
| 14 | Poplar | 9.5 (Kheshgi et al. 2000) | 19.8 | 0.98 | 0.36 |
| 14 | Wood grown sustainably in Western Russia (Nabuurs and Lioubimov 2000) | 1.4 | 19.8 | 0.95 (Reijnders and Huijbregts 2003) | 0.05 |

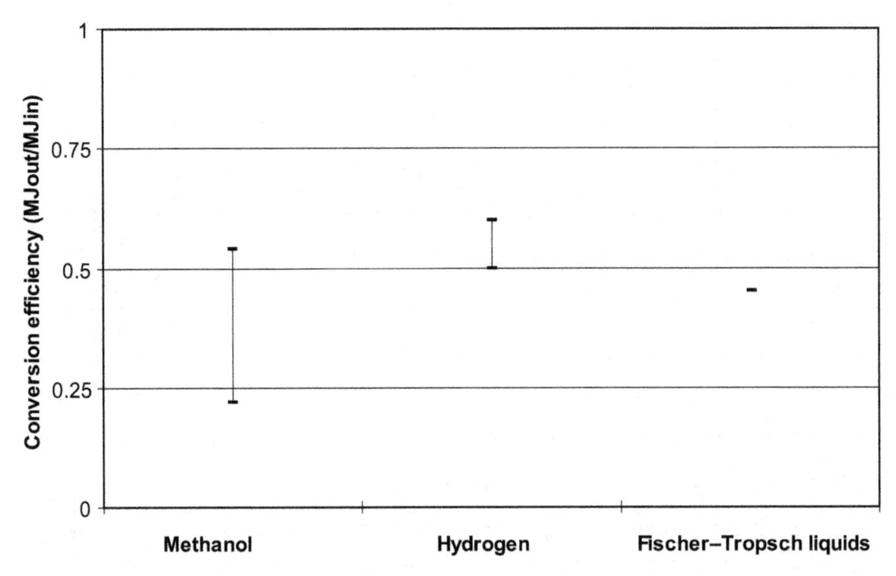


**Fig. 2.1** Estimated biomass to transport biofuel conversion efficiencies via synthesis gas (Chum and Overend 2001; Ptasinki et al. 2002; Iwasaki 2003; Faaij 2006)

verted into the biochemicals needed is used to power the production from biomass of specific biochemicals. It may also be that a part of the original biomass emerges from processing as waste. Furthermore, in many processes generating biochemicals from biomass, there is an input of fossil fuels that is to be taken into account when determining overall conversion efficiencies. Figure 2.1 gives estimated efficiencies for some conversions of biomass into transport biofuels.

Table 2.2 shows solar conversion efficiencies for a number of biofuels from terrestrial plants. In this case, the allocation has been done on the basis of energy content of marketable products.

The efficiencies in the last column of Table 2.2 are typically lower than the efficiencies shown in Table 2.1. Most of them are below 0.2%. For ethanol from European wheat starch, the efficiency is 0.024–0.03%, and for biodiesel from European rapeseed, it is approximately 0.034%. Apart from *Jatropha*, which has quite an uncertain conversion efficiency, the best efficiency in Table 2.2 is for ethanol from switchgrass, with ethanol from sugar cane coming second. However, it was assumed in this table that in the case of sugar cane, only sugar is to be converted into ethanol. If also a substantial part of the lignocellulosic aboveground biomass of sugar cane is converted into ethanol, sugar cane may as efficient as or better than switchgrass.

**Table 2.2** Efficiency of solar energy conversion into specific biochemicals, when corrected for fossil fuel inputs and when allocation is based on prices

| Crop/Process | Insolation (MJ day$^{-1}$/m$^2$) | Fuel | Yield of biofuel ha$^{-1}$ (Mg/year) | Heat of combustion of biofuel (MJ/kg) | MJ biofuel – MJ fossil fuel input/MJ biofuel | MJ biofuel – MJ fossil/MJ crop | Solar energy conversion efficiency (%) |
|---|---|---|---|---|---|---|---|
| Oil palm | 19 | Palm kernel oil | | 40 | 0.7–0.9 (Reijnders and Huijbregts 2008a; de Vries 2008) | | ~ 0.15 (Reijnders 2008) |
| Sugar cane | 19 | Ethanol | | 26.4 | | | 0.16 (Kheshgi et al. 2000) |
| Wheat (Europe) | 14 | Ethanol | 1.65 (from starch) | 26.4 | 0.2–0.25 (Somerville 2007) | | 0.024–0.03 |
| Switchgrass to be processed to ethanol | 14 | Ethanol | 15 | 26.4 | 0.25 (Fleming et al. 2006; Champagne 2007) | | 0.2 |
| *Jatropha* | 19 | Oil | 0.65–5 (Fairless 2007; van Eijck and Romijn 2008) | 40 | 0.9 or lower | | 0.035–< 0.26 |
| Switchgrass | 14 | Fischer–Tropsch diesel | 15 | 44 | | 0.14 (Somerville 2007) | 0.18 |
| Wood from trees | 14 | Methanol generated by dry distillation | 0.04 (Reinharz 1985) | 19.8 | | | 0.0015 |
| Wood (Europe) | 14 | Fischer–Tropsch diesel | 1.4 | 44 | | 0.12–0.33 (Huber et al. 2006) | 0.014–0.039 |
| Rapeseed | 14 | Biodiesel | 1.15 (oil, before transesterification) | | | 0.4 (Reijnders and Huijbregts 2008b) | 0.034 |

### *2.4.3 Biofuels from Algae and Aquatic Macrophytes*

As also pointed out in Chap. 1, estimates have been made of the maximum efficiency for the conversion of incident sunlight into biomass by algae. These vary between 5.5 and 11.6% (Heaton et al. 2008b; Vasudevan and Briggs 2008). Several authors have suggested that algal transport biofuels can beat terrestrial transport biofuels in the conversion of solar energy to transport biofuel by at least one order of magnitude (e.g. Chisti 2007; Chisti 2008a; Groom et al. 2008; Nowak 2008; Li et al. 2008). Here we will survey the suggestions that have been made for producing transport biofuels from algae and aquatic macrophytes and what is known about the solar energy conversion efficiency of such biofuels.

#### Transport Biofuels from Marine Aquatic Biomass

As pointed out in Chap. 1, a variety of options for producing biofuels from marine biomass have been suggested, such as biofuels from *Macrocystis pyrifera* or giant kelp (Wilcox 1982; Bungay 2004), *Laminaria* (Horn et al. 2000; Chopin et al. 2001) and *Dunaliella* (Ben-Amotz et al. 1982). *Dunaliella* has been found more suitable to cultivation in open ponds (Joint et al. 2002; Ugwu et al. 2008). As to *Macrocystis pyrifera,* it seems doubtful whether the energy balance for biofuel can be positive (Bungay 2004).

Near-shore cultivation of macroalgae is substantial (Neushul and Wang 2000; Wikfors and Ohno 2001; Chopin et al. 2001; Critchley et al. 2006; Troell et al. 2006). For *Gracilaria* in Taiwanese coastal waters, average yields of 4 Mg $ha^{-1}$ $year^{-1}$ (dry weight) have been reported (van der Meer 1983). Yields of commercial *Eucheuma* cultivation in the Philippines, Indonesia and Kiribati are about 6 Mg (dry weight) $ha^{-1}$ $year^{-1}$ (Ask and Azanza 2002). Such yields suggest relatively low solar energy conversion efficiencies if compared with cultivated terrestrial plants (see Table 2.1). Cultivation is vulnerable to invasions of competing algae and herbivores, and major interventions may be necessary to limit losses in such cases (Buschmann et al. 2001; Ask and Azanza 2002; Neill et al. 2006). As pointed out in Chap. 1, prices for cultivated macroalgae are high, and thus it is hard to see the emergence of a practical large-scale biomass-from-the-sea-for-transport-fuel scheme based on macroalgae cultivation (Neushul and Badash 1998; Buschmann et al. 2001).

#### Microalgal Biomass in Ponds and Bioreactors

Most proposals for microalgal biofuels from open ponds or bioreactors focus on biodiesel made from algal oil (Scragg et al. 2002; Chisti 2007; Huntley and Redalje 2007; Wijffels 2008; www.oilgae.com; Liu et al. 2008). However, there have, for instance, also been proposals to convert algal biomass into methanol via synthesis gas or into bio-oil via pyrolysis (Hirano et al. 1998; Sawayama et al. 1999). The alga *Botryococcus braunii* has been looked into, in view of its ability to produce substantial amounts of hydrocarbons, which may be turned into transport biofuels by catalytic cracking (Bachofen 1982; Banerjee et al. 2002). As pointed out in Chap. 1, current strains of this microalga are slow growing, which has not been conducive to its application (Banerjee et al. 2002).

Of the microalgae commercially grown in open ponds, *Spirulina* apparently has the best yields per hectare per year in commercial cultivation (Belasco 1997). Maximum productivities in open ponds are achieved under tropical or subtropical conditions (Jiménez et al. 2003). Yields currently obtained in industrial facilities for the cultivation of *Spirulina* located in these regions range from 10 to 30 Mg dry biomass per hectare per year (Vonshak and Richmond 1988; Jiménez et al. 2003). Low yields of, for example, *Spirulina* may however occur due to, for example, phage infections and rainfall conducive to the growth of unfavourable organisms (Shimamatsu 2004). For instance, Li and Qi (1997) reported that the 80 Chinese *Spirulina* production plants had production on average of 3.5 $Mg\,ha^{-1}\,year^{-1}$.

It may be that in the future, microalgal yields from raceway ponds may be increased over current levels, for instance through improving photosynthetic activity by minimizing light harvesting chlorophyll antenna size (Neidhardt et al. 1998; Mussgnug et al. 2007). On the other hand, a focus on algal lipids for transport biofuel production may well lead to biomass yield limitations, because nutrient limitations are conducive to high lipid contents but not to maximizing biomass yield (Wijffels 2008; Liu et al. 2008).

Hirano et al. (1998) studied *Spirulina* production and processing to supply methanol (via synthesis gas) and assumed a yield of approximately 110 $Mg\,ha^{-1}\,year^{-1}$. When both fossil fuel inputs in infrastructure and operation are considered, this would correspond with an overall solar energy to biofuel conversion efficiency of about 0.12%.

Actual yearly yields much exceeding 30 $Mg\,ha^{-1}\,year^{-1}$ have been claimed for microalgae growing in water that has been saturated in $CO_2$ (Kheshgi et al. 2000; Wang et al. 2008). Algal ponds that are to be saturated in $CO_2$ have been proposed to capture the $CO_2$ of power plants (Kheshgi et al. 2000). Also, closed bioreactors have been proposed for algal capture of $CO_2$ from power plants (Skjånes et al. 2007). The efficiency of algal $CO_2$ capture in open ponds has been estimated to be in the order of 30% (Benemann 1993; Kadam 2002), whereas an efficiency of 40% has been suggested for algae in photobioreactors (Ono and Cuello 2006). Whether such percentages can be achieved is not certain. Yields from open ponds saturated with $CO_2$ have proved disappointing, and maintaining desired algal cultures in such ponds has turned out to be difficult (Benemann et al. 2003). There is also the matter of the efficiency of $CO_2$ sequestration by algae. The suggested efficiency for photobioreactors of 40% is, for instance, higher than efficiencies so far reported by Hsueh et al. (2007) and Jacob-Lopes et al. (2008) for flue gases with high concentrations of $CO_2$ handled by photobioreactors. Moreover, the latter efficiencies were achieved under good irradiation, whereas the $CO_2$ emission of power plants may also occur at night and when solar irradiation is poor. $CO_2$ capture and sequestration (CCS) in aquifers or abandoned natural gas or oil fields would be able to reduce the emission of power plants with an efficiency of about 90% (Odeh and Cockerill 2008). Thus, whether the application of $CO_2$ capture by algae will be important in the future depends to a large extent on the emission requirements for such plants.

Microalgal yields from closed bioreactors subject to solar irradiation may be much higher than from current commercial open ponds (Eriksen 2008). For the production of algal oil, a value of about $16\,\mathrm{Mg\,ha^{-1}\,year^{-1}}$, has been suggested as ‘possible with state of the art technology’ in closed systems (Wijffels 2008). However, growing algae aiming at high outputs in bioreactors requires large inputs of energy for building the reactors and for nutrients and intensive mixing. It has been estimated that this could lead to a negative energy balance for flat panel bioreactors and an even more negative energy balance for tubular bioreactors (Wijffels 2008).

### $H_2$ Produced by Microalgae

The use of a variety of algae has been considered because of their direct and indirect biocatalytic effect on the splitting of water in $H_2$ and $O_2$ (Melis and Happe 2001; Hallenbeck and Benemann 2002; Nath and Das 2004; Savage et al. 2008). In spite of a nearly 70-year history of research, actual production of $H_2$ by algal systems is still very low, about 2 g of $H_2$ per square metre of culture area per day (Melis and Happe 2001), and $H_2$ has to be withdrawn continually as the overall conversion of glucose into $H_2$ is energetically only slightly favourable to $H_2$ (Savage et al. 2008). At realistic solar irradiation, solar conversion efficiencies in optimized systems for direct and indirect biophotolysis seem to be in the order of 1% or lower, when pure cultures can be maintained (Hallenbeck and Benemann 2002; Yoon et al. 2006; Rupprecht et al. 2006; Burgess and Fernández-Velasco 2007). And as pointed out before, the cumulative energy demand for algal $H_2$ production is probably of the same order of magnitude as the energetic output, when the solar energy conversion efficiency does not exceed 1% (Burgess and Fernándo-Velasco 2007).

### Freshwater Macrophytes

The best-studied macrophyte is the water hyacinth (*Eichhornia crassipes*) (Gassmann et al. 2006; Gunnarsson and Petersen 2007). It has been found to produce up to $140\,\mathrm{Mg\,ha^{-1}\,year^{-1}}$ of biomass (dry weight) (Gunnarsson and Petersen 2007). Two energetic applications of *Eichhornia crassipes* which may produce transport biofuels have been studied. The first is ethanol production from hemicellulose present in water hyacinths. A yield of 0.14–0.17 (g ethanol) (g dry weight)$^{-1}$ has been reported (Mishima et al. 2008). However, studies of the overall energy efficiency of the production of ethanol from the water hyacinth have so far suggested that the energy balance is negative (Gunnarsson and Petersen 2007). An alternative option is the anaerobic conversion of water hyacinth biomass into $CH_4$. Though the feasibility thereof has been demonstrated, the process is complicated, among other things by the floating behaviour of water-hyacinth-derived material (Malik 2007). Moreover, the water hyacinth is very effective in adsorbing pollutants (Gunnarsson and Petersen 2007; Malik 2007), and these may interfere with, for example, the sustainable use of residuals ('digestate') remaining after anaerobic conversion. More limited research has been done regarding another invasive macrophyte: water lettuce (*Pistia stratiotes* L.), which has growth characteristics similar to water hyacinth (Mishima et al. 2008). The yield of ethanol from hemicellulose conversion is 0.15–0.16 g per gram of dry weight (Mishima et al. 2008); no study has been found regarding the overall energy efficiency of this conversion.

## 2.5 Solar Conversion Efficiencies of Physical Methods

Besides biological processes, there are also physical conversion processes for solar energy. Efficiencies for a number of physical methods of converting solar radiation into heat, $H_2$ or electricity are in Table 2.3. It can be seen that solar conversion efficiencies of photovoltaic cells are much higher than the conversion efficiencies for the transport biofuels in Table 2.2.

**Table 2.3** Efficiencies for the conversion of solar radiation to electricity or heat

| Type of conversion | Output | Conversion efficiency | Correction factor for fossil fuel input into conversion apparatus (MJ output – fossil fuel input/MJ output) | Overall energy efficiency (%) |
|---|---|---|---|---|
| Photovoltaic silicon (Mohr et al. 2007; Fthenakis et al. 2008) | Electricity | ∼14 | 0.75–0.8 | ∼10.5–12 |
| Hybrid photovoltaic silicon/ collector | Electricity/ heat | 15% (electricity) +40% heat (He et al. 2006; Tripanagnostopoulos et al. 2006) | 0.9–0.95 | 49.5–52 |
| Photovoltaic III–V | Electricity | 15–30 (Green et al. 2003) | 0.8–0.9 (dependent on insolation) (Meijer et al. 2003; Mohr et al. 2007) | 12–27 |
| Solar thermal electricity turbine | Electricity | 10–28% (Mancini et al. 1994) | 0.93 (Norton et al. 1998) | 9.5–26.5 |

## 2.6 Overall Energy Efficiencies in Performing Work

Table 2.4, finally, shows overall estimated conversion efficiencies for solar irradiation to car kilometres, corrected for the input of fossil fuels, which are calculated by

$$TSCE_{x,i} = SCE_x \cdot CE_{x,i}$$

where $TSCE_{x,i}$ is the transport solar energy conversion efficiency (%) and $CE_{x,i}$ the efficiency drive train of transport option $i$ derived from biofuel type $x$ (%).

According to the estimates in Table 2.4 regarding seed-to-wheel solar energy conversion efficiency, ethanol from sugar cane outperforms ethanol from European wheat by about a factor of five to ten, and biodiesel from European rapeseed by about a factor of two to three. Electrical traction from lignocellulosic biomass, how-

**Table 2.4** Overall efficiencies for the conversion of solar energy to car kilometres

| Type of energy supply | Conversion efficiency solar radiation to automotive power source, corrected for fossil fuel inputs (%); see Tables 2.2 and 2.3 | Efficiency drive train (%) | Overall efficiency energy storage (%) | Overall efficiency conversion solar radiation to automotive kilometres (%) |
|---|---|---|---|---|
| Ethanol from sugar cane (Brazil) for Otto motor | 0.16 | 16–22 (Colella et al. 2005; Crabtree et al. 2004) | | 0.026–0.035 |
| Ethanol from wheat (Europe) for Otto motor | 0.024–0.03 | 16–22 (Colella et al. 2005; Crabtree et al. 2004) | | 0.0038–0.0066 |
| Biodiesel from rapeseed (Europe) for diesel motor | 0.034 | 29 (www.eere.energy.gov/vehiclesandfuels) | | 0.010 |
| Electricity from lignocellulosic biomass (switchgrass) for electromotor | 0.48 | 90–97 (Ahluwalia et al. 2005; Colella et al. 2005) | 41–90 (Rydh and Sandén 2005) | 0.18–0.42 |
| Electricity from solar cells for electromotor | 10.5–12 | 90–97 (Ahluwalia et al. 2005; Colella et al. 2005) | 41–90 (Rydh and Sandén 2005) | 3.9–10.5 |

ever, in turn outperforms ethanol from sugar cane by roughly a factor of two to four. The relatively high efficiency of using biomass for electricity production has also been noted by other authors, such as Zhang et al. (2007). All biomass-based automotive power is, however, far less efficient than electricity from solar cells that is stored for use in electrical traction. This way of powering motor cars is roughly at least two orders of magnitude better than ethanol from Brazilian sugar cane and three orders of magnitude better than ethanol from European wheat. In calculating the values for Table 2.4, it has been assumed that solar cells and the plug-in facility for cars are in the same region. When distances are large or conversion to $H_2$ is necessary for long distance transport, the efficiency will be lower than indicated in Table 2.4 because of transport-linked losses. For instance, an estimate has been made regarding the life cycle emission of greenhouse gases linked to electrolysis powered by concentrated solar power (CSP) in the Sahara, liquefaction of $H_2$ and transport to, and distribution of, hydrogen in Western Europe. In such a case, a reduction of the life cycle efficiency by somewhat less than 10% has been found (Ros et al. 2009). Such a reduction applied to electricity from solar cells (last row of Table 2.4) would reduce the overall efficiency in the last column to approximately 3.5–9.4%.

A lesson from this chapter is that conversions lead to substantial reductions in solar conversion efficiency. In Chap. 1, quite a number of proposals have been summarized that rely on such conversions. Examples are: the conversion of methane (from the anaerobic conversion of biomass) to methanol, the conversion of lipids and ethanol to hydrocarbons or $H_2$ and the conversion of methanol to hydrocarbons. As the starting products may in principle be used directly as transport biofuels, there is good reason to be sceptical about such sequential conversions in view of the negative impact that they have on the overall solar energy conversion efficiency.

The data presented in this chapter allow for estimates of the ability of biofuels to energetically displace fossil fuels. It appears that in this respect, palm oil and ethanol from sugar cane do much better, especially when processing is powered by harvest residues, than rapeseed oil or ethanol from corn or wheat, as produced in industrialized countries. It should be noted, though, that the ability to energetically displace fossil fuels may be at variance with their ability to do so in the economy. The latter is strongly impacted by prices and government policy. An interesting illustration thereof concerns the use of corn-derived ethanol in US gasoline, which has mainly been by E10 fuels, containing 10% ethanol and 90% conventional gasoline. The use of E10 fuels has been stimulated by a federal excise tax which in recent years led to E10 gasoline being cheaper than conventional gasoline (Tyner 2008; Vedenov and Wetzstein 2008), which in turn had an upward effect on the overall consumption of gasoline, thereby partly negating the downward effect of ethanol use on the consumption of conventional gasoline (Vedenov and Wetzstein 2008).

The data in this chapter also allow for estimates of land requirements linked to a large-scale displacement of fossil transport fuels by biofuels. This may be illustrated by the following back-of-the-envelope calculation. As explained in Chap. 1, mineral oil is the dominating fossil fuel for powering transport, and about 60% of all crude oil is used for this transport. Let us suppose that all mineral oil that is currently used as an input in worldwide transport were to be replaced by vegetable oil. Corrected for the difference in lower heating value between crude oil and vegetable oil (see Table 1.2) and the cumulative fossil fuel input into vegetable oil (estimated here at 40% of the energetic value of vegetable oil), this would require an increase of vegetable oil production by about a factor of 37.5. Part of this increase may be met with the increase of yields per hectare. Estimates made for the 23 most important food crops suggest that such an increase may range from 0.63–1.76% $year^{-1}$ for developing countries and from 0.59–0.79% $year^{-1}$ for developed countries up to 2050 (Balmford et al. 2005), to a large extent by intensification of cropping (Tilman et al. 2001). Using intermediate values, this would allow for an increase in yield by a factor of approximately 1.75 for developing countries and by a factor of approximately 1.42 for developed countries between 2000 and 2050 (Balmford et al. 2005), far below the factor of 37.5 needed to displace all mineral oil by vegetable oil. Moreover, it may well be that the average productivity of additional land is lower than that of land currently in use. Thus, even if yield increases in the future would be much larger than currently estimated, there would seem no way around large additional land requirements linked to large-scale displacement of fossil fuels by biofuels. Current policy targets are estimated to require between 55 and 166 million ha (Mha) (Renewable Fuels Agency 2008).

Moreover, expanding transport may well lead to even larger land claims in the future. Gurgel et al. (2007) studied an expansion of the production of cellulosic biofuel to supply up to 368 EJ in 2100. This, according to their scenario, would require about $2.5 \times 10^3$ Mha, an amount greater than any other land cover category. For comparison: worldwide, current cropland is about $1.6 \times 10^3$ Mha, and the land area that is currently considered fit for additional cropland is estimated at between 400 and approximately $1.2 \times 10^3$ Mha (Renewable Fuels Agency 2008).

## References


Aaronson S, Dubinsky Z (1982) Mass production of microalgae. Experientia 38:36–40

Abrahamson LP, Robison DJ, Volk TA, White EH, Neuhauser EF, Benjamin WH, Peterson JM (1998) Sustainability and environmental issues associated with willow bioenergy development in New York (U.S.A.). Biomass Bioenerg 15:17–22

Ahluwalia RK, Wang X, Rousseau A (2005) Fuel economy of hybrid fuel-cell vehicles. J Power Sources 152:233–244

Akkerman I, Janssen M, Rocha J, Wijffels RH (2002) Photobiological hydrogen production: photochemical efficiency and bioreactor design. Int J Hydrogen Energ 27:1195–1208

Ask EI, Azanza RV (2002) Advances in cultivation technology of commercial eucheumatoid species: a review with suggestions for future research. Aquaculture 206:257–277

Bachofen R (1982) The production of hydrocarbons by *Botryococcus braunii*. Experientia 38: 47–49

Balmford A, Green RE, Scharlemann JPW (2005) Sparing land for nature: exploring the potential impact of changes in agricultural yield on the area needed for crop production. Glob Change Biol 11:1594–1605

Banerjee A, Sharma R, Chisti Y, Banerjee UC (2002) *Botryococcus braunii*: a renewable source of hydrocarbons and other chemicals. Crit Rev Biotechnol 22:245–279

Bastiaanssen WGM, Ali S (2003) A new crop yield forecasting model based on satellite measurements applied across the Indus Basin, Pakistan. Agr Ecosyst Environ 94:321–340

Bastianoni S, Coppola F, Tiezzi E, Colacevich A, Borghini F, Focardi S (2008) Biofuel potential production from the Orbetello lagoon macroalgae: a comparison with sunflower feedstock. Biomass Bioenerg 32:619–628

Belasco W (1997) Algae burgers for a hungry world? The rise and fall of Chlorella cuisine. Technol Cult 38:608–634

Ben-Amotz A, Sussman I, Avron M (1982) Glycerol production by *Dunaliella*. Experientia 38: 49–52

Benemann JR (1993) Utilization of carbon dioxide from fossil fuel-burning power plants with biological systems. Energ Convers Manage 34:999–1004

Benemann JR, Van Olst JC, Massingill MJ, Carlberg JA, Weissman JC, Brune DE (2003) The controlled eutrophication process: using microalgae for $CO_2$ utilization and agricultural fertilizer recycling. In: Greenhouse gas control technologies – 6th international conference, pp 1433–1438

Bhattacharya SC, Abdul Salam P, Pham HL, Ravindranath NH (2003) Sustainable biomass production for energy in selected Asian countries. Biomass Bioenerg 25:471–482

Bolton JR (1996) Solar photoproduction of hydrogen: a review. Sol Energy 57:37–50

Braunack MV, McGarry D, Sugar Yield Decline Joint Venture (2006) Traffic control and tillage strategies for harvesting and planting of sugarcane (Saccharum officinarum) in Australia. Soil Till Res 89:86–102

Buckley JC, Schwarz PM (2003) Renewable energy from gasification of manure: an innovative technology in search of fertile policy. Environ Monit Assessment 84:111–127

Bungay HR (2004) Confessions of a bioenergy advocate. Trends Biotechnol 22:67–71

Burgess G, and Fernández-Velasco JG (2007) Materials, operational energy inputs, and net energy ratio for photobiological hydrogen production. Int J Hydrogen Energ 32:1225–1234

Buschmann AH, Correa JA, Westermeier R, Hernández-González MdC, Norambuena R (2001) Red algal farming in Chile: a review. Aquaculture 194:203–220

Butterworth B (2006) Biofuels from waste: closed loop sustainable fuel production. Refocus 7(3):60–61

Çelik AN (2006) Analysis of Ankara's exposure to solar radiation: evaluation of distributional parameters using long-term hourly measured global solar radiation data. Turkish J Eng Env Sci 30:115–126

Champagne P (2007) Feasibility of producing bio-ethanol from waste residues: a Canadian perspective. Resour Conserv Recy 50:211–230

Chisti Y (2007) Biodiesel from microalgae. Biotechnol Adv 25:294–306

Chisti Y (2008a) Biodiesel from microalgae beats bioethanol. Trends Biotechnol 26:126–131

Chisti Y (2008b) Response to Reijnders: do biofuels from microalgae beat biofuels from terrestrial plants? Trends Biotechnol 26:351–352

Chopin T, Buschmann AH, Halling C, Troell M, Kautsky N, Neori A, Kraemer GP, Zertuche-González JA, Yarish C, Neefus C (2001) Integrating seaweeds into marine aquaculture systems: a key towards sustainability. J Phycol 37:975–986

Christian DG, Riche AB, Yates NE (2008) Growth, yield and mineral content of *Miscanthus x giganteus* grown as a biofuel for 14 successive harvests. Ind Crop Prod 28:320–327

Chum HL, Overend RP (2001) Biomass and renewable fuels. Fuel Process Technol 71:187–195

Colella WG, Jacobson MZ, Golden DM (2005) Switching to a U.S. hydrogen fuel cell vehicle fleet: the resultant change in emissions, energy use, and greenhouse gases. J Power Sources 150:150–181

Coombs J, Lynch JM, Levy JF, Gascoigne JA (1987) EEC resources and strategies [and discussion]. Philos T R Soc S-A 321:405–422

Crabtree GW, Dresselhaus MS, Buchanan MV (2004) The hydrogen economy. Phys Today December:39–44

Critchley AT, Ohno M, Largo DB (eds) (2006) World seaweed resources: an authoritative reference system. ETI Information Services, Wokingham (UK)

Demirbaş A (2007) Progress and recent trends in biofuels. Prog Energ Combust 33:1–18
de Vries SC (2008) The bio-fuel debate and fossil energy use in palm oil production: a critique of Reijnders and Huijbregts 2007. J Clean Prod 16:1926–1927
Dias de Oliveira ME, Vaughan BE, Rykiel EJ (2005) Ethanol as fuel: energy, carbon dioxide balances, and ecological footprint. BioScience 55:593–602
Dismukes GC, Carrieri D, Bennette N, Ananyev GM, Posewitz MC (2008) Aquatic phototrophs: efficient alternatives to land-based crops for biofuels. Curr Opin Biotechnol 19:235–240
Eickhout B, van den Born GJ, Notenboom J, van Oorschot M, Ros JPM, van Vuuren DP, Westhoek HJ (2008) Local and global consequences of the EU renewable directive for biofuels. Milieu en Natuur Planbureau Bilthoven. http://www.mnp.nl
Ekvall T, Finnveden G (2001) Allocation in ISO 14041 – a critical review. J Clean Prod 9:197–208
El Bassam N (1998) $C_3$ and $C_4$ plant species as energy sources and their potential impact on environment and climate. Renew Energ 15:205–210
Elobeid A, Hart C (2007) Ethanol expansion in the food versus fuel debate: how will developing countries fare? J Agric Food Ind Organ 5(2):article 6. http://www.bepress.com/jafio/vol5/iss2/art6/
Eriksen NT (2008) The technology of microalgal culturing. Biotechnol Lett 30:1525–1536. doi:10.1007/s1052900897403
Faaij A (2006) Modern biomass conversion technologies. Mitigation Adaptation Strategies Global Change 11:335–367
Fairless D (2007) Biofuel: the little shrub that could – maybe. Nature 449:652–655
Falkowski PG (1994) The role of phytoplankton photosynthesis in the biochemical cycles. Photosynth Res 39:235–258
Fleming JS, Habibi S, MacLean H (2006) Investigating the sustainability of lignocellulose-derived fuels for light-duty vehicles. Transport Res D-Tr E 11:146–159
Fthenakis VM, Kim HC, Alsema E (2008) Emissions from photovoltaic life cycles. Environ Sci Technol 42:2168–2174
Gassmann A, Cock MJW, Shaw R, Evans HC (2006) The potential for biological control of invasive alien aquatic weeds in Europe: a review. Hydrobiologia 570:217–222
Goldemberg J, Teixeira Coelho S (2004) Renewable energy – traditional biomass vs. modern biomass. Energ Policy 32:711–714
Goldemberg J, Teixeira Coelho S, Guardabassi P (2008) The sustainability of ethanol production from sugarcane. Energ Policy 36:2086–2097
Gordon JM, Polle JEW (2007) Ultrahigh bioproductivity from algae. Appl Microbiol Biotechnol 76:969–975
Green MA, Emery K, King DL, Igari S, Warta W (2003) Solar efficiency tables (version 22). Prog Photovoltaics Res Appl 11:347–352
Groom MJ, Gray EM, Townsend PA (2008) Biofuels and biodiversity: principles for creating better policies for biofuel production. Conserv Biol 22:602–609
Gunnarsson CC, Petersen CM (2007) Water hyacinths as a resource in agriculture and energy production: a literature review. Waste Manage 27:117–129
Gurgel A, Reilly JM, Paltsev S (2007) Potential land use implications of a global biofuels industry. J Agric Food Ind Organ 5(2):article 9. http://www.bepress.com/jafio/vol5/iss2/art9/
Hall DO (1982) Solar energy through biology: fuels from biomass. Experientia 38:3–10
Hallenbeck PC, Benemann JR (2002) Biological hydrogen production; fundamentals and limiting processes. Int J Hydrogen Energ 27:1185–1193
Hamelinck CN, Faaij APC (2002) Future prospects for production of methanol and hydrogen from biomass. J Power Sources 111:1–22
Hammerschlag R (2006) Ethanol's energy return on investment: a survey of the literature 1990–present. Environ Sci Technol 40:1744–1750
He W, Chow T, Ji J, Lu J, Pei G, Chan L (2006) Hybrid photovoltaic and thermal solar-collector designed for natural circulation of water. Appl Energ 83:199–210
Heaton EA, Dohleman FG, Long SP (2008a) Meeting US biofuel goals with less land: the potential of *Miscanthus*. Glob Change Biol 14:2000–2014

Heaton EA, Flavell RB, Mascia PN, Thomas SR, Dohleman FG, Long SP (2008b) Herbaceous energy crop development: recent progress and future prospects. Curr Opin Biotechnol 19: 202–209

Hill J, Nelson E, Tilman D, Polasky S, Tiffany D (2006) Environmental, economic, and energetic costs and benefits of biodiesel and ethanol fuels. P Natl Acad Sci USA 103:11206–11210

Hirano A, Hon-Nami K, Kunito S, Hada M, Ogushi Y (1998) Temperature effect on continuous gasification of microalgal biomass: theoretical yield of methanol production and its energy balance. Catal Today 45:399–404

Horn SJ, Aasen IM, Østgaard K (2000) Ethanol production from seaweed extract. J Ind Microbiol Biotechnol 25:249–254

Hsueh HT, Chu H, Yu ST (2007) A batch study on the bio-fixation of carbon dioxide in the absorbed solution from a chemical wet scrubber by hot spring marine algae. Chemosphere 66:878–886

Huber GW, Iborra S, Corma A (2006) Synthesis of transportation fuels from biomass: chemistry, catalysts, and engineering. Chem Rev 106:4044–4098

Huntley ME, Redalje DG (2007) $CO_2$ mitigation and renewable oil from photosynthetic microbes: a new appraisal. Mitigation Adaptation Strategies Global Change 12:573–608

Iwasaki W (2003) A consideration of the economic efficiency of hydrogen production from biomass. Int J Hydrogen Energ 28:939–944

Jacob-Lopes E, Scoparo CHG, Franco TT (2008) Rates of $CO_2$ removal by *Aphanothece microscopica Nägeli* in tubular photobioreactors. Chem Eng Process 47:1365–1373

Jiménez C, Cossio BR, Niell FX (2003) Relationship between physicochemical variables and productivity in open ponds for the production of *Spirulina*: a predictive model of algal yield. Aquaculture 221:331–345

Joint I, Henriksen P, Garde K, Riemann B (2002) Primary production, nutrient assimilation and microzooplankton grazing along a hypersaline gradient. FEMS Microbiol Ecol 39:245–257

Kadam KL (2002) Environmental implications of power generation via coal-microalgae cofiring. Energy 27:905–922

Karp A, Shield I (2008) Bioenergy from plants and the sustainable yield challenge. New Phytol 179:15–32

Keoleian GA, Volk TA (2005) Renewable energy from willow biomass crops: life cycle energy, environmental and economic performance. Crit Rev Plant Sci 24:385–406

Kheshgi HS, Prince RC, Marland G (2000) The potential of biomass fuels in the context of global climate change: focus on transportation fuels. Annu Rev Energ Env 25:199–244

Kim S, Dale BE (2005) Life cycle assessment of various cropping systems utilized for producing biofuels: bioethanol and biodiesel. Biomass Bioenerg 29:426–439

Kim S, Dale BE (2008) Life cycle assessment of fuel ethanol derived from corn grain via dry milling. Bioresour Technol 99:5250–5260

Koizumi T, Ohga K (2007) Biofuels policies in Asian countries: impact of the expanded biofuels programs on world agricultural markets. J Agric Food Ind Organ 5(2):article 8. http://www.bepress.com/jafio/vol5/iss2/art8/

Krotscheck C, König F, Obernberger I (2000) Ecological assessment of integrated bioenergy systems using the Sustainable Process Index. Biomass Bioenerg 18:341–368

Lemus R, Lal R (2005) Bioenergy crops and carbon sequestration. Crit Rev Plant Sci 24:1–21

Lewandowski I, Schmidt U (2006) Nitrogen, energy and land use efficiencies of miscanthus, reed canary grass and triticale as determined by the boundary line approach. Agr Ecosyst Environ 112:335–346

Lewis NS, Nocera DG (2006) Powering the planet: chemical challenges in solar energy utilization. P Natl Acad Sci USA 43:15729–15735

Li D, Qi Y (1997) *Spirulina* industry in China: present status and future prospects. J Appl Phycol 9:25–28

Li Y, Horsman M, Wu N, Lan CQ, Dubois-Calero N (2008) Biofuels from microalgae. Biotechnol Prog. doi:10.1021/bp070371k

Licht S, Wang B, Mukerji S, Soga T, Umeno M, Tributsch H (2001) Over 18% solar energy conversion to generation of hydrogen fuel; theory and experiment for efficient solar water splitting. Int J Hydrogen Energ 26:653–659

Liu Z, Wang G, Zhou B (2008) Effect of iron on growth and lipid accumulation in *Chlorella vulgaris*. Bioresour Technol 99:4717–4722

Long SP, Zhu X, Naidu SL, Ort DR (2006) Can improvement in photosynthesis increase crop yields? Plant Cell Environ 29:315–330

Macedo IC, Seabra JEA, Silva JEAR (2008) Greenhouse gases emissions in the production and use of ethanol from sugarcane in Brazil: the 2005/2006 averages and a prediction for 2020. Biomass Bioenerg 32:582–595

Malhi SS, Lemke R, Wang ZH, Chabra BS (2006) Tillage, nitrogen and crop residue effects on crop yield, nutrient uptake, soil quality, and greenhouse gas emissions. Soil Till Res 90:171–183

Malik A (2007) Environmental challenge *vis a vis* opportunity: the case of water hyacinth. Environ Int 33:122–138

Mancini TR, Chavez JM, Kolb GJ (1994) Solar thermal power today and tomorrow. Mech Eng 116(8):74–79

Meijer A, Huijbregts MAJ, Schermer JJ, Reijnders L (2003) Life-cycle assessment of photovoltaic modules: comparison of mc-Si, InGaP and InGaP/mc-Si solar modules. Prog Photovoltaics Res Appl 11:275–287

Melis A, Happe T (2001) Hydrogen production: green algae as a source of energy. Plant Physiol 127:740–748

Mezhunts BK, Givens DJ (2004) The productivity and energy value of mountain grasslands in Armenia. J Nat Sci Natl Acad Sci Repub Armenia 1:13–19

Mishima D, Kuniki M, Sei K, Soda S, Ike M, Fujita M (2008) Ethanol production from candidate energy crops: water hyacinth (*Eichhornia crassipes*) and water lettuce (*Pistia stratiotes* L.). Bioresour Technol 99:2495–2500

Mohr NJ, Schermer JJ, Huijbregts MAJ, Meijer A, Reijnders L (2007) Life cycle assessment of thin-film GaAs and GaInP/GaAs solar modules. Prog Photovoltaics Res Appl 15:163–179

Morand P, Merceron M (2005) Microalgal population and sustainability. J Coastal Res 21: 1009–1020

Murphy JD, Power NM (2008) How can we improve the energy balance of ethanol production from wheat. Fuel 87:1799–1806

Mussgnug JH, Thomas-Hall S, Rupprecht J, Foo A, Klassen V, McDowall A, Schenk PM, Kruse O, Hankamer B (2007) Engineering photosynthetic light capture: impacts on improved solar energy to biomass conversion. Plant Biotechnol J 5:802–814

Nabuurs GJ, Lioubimov AV (2000) Future development of the Leningrad region forests under nature-oriented forest management. Forest Ecol Manag 130:235–251

Nath K, Das D (2004) Improvement of fermentative hydrogen production: various approaches. Appl Microbiol Biotechnol 65:520–529

Neidhardt J, Benemann JR, Zhang L, Melis A (1998) Photosystem-II repair and chloroplast recovery from irradiance stress: relationship between chronic photoinhibition, light-harvesting chlorophyll antenna size and photosynthetic productivity of Dunaliella salina (green algae). Photosynth Res 56:175–184

Neill PE, Alcalde O, Faugeron S, Navarrete SA, Correa JA (2006) Invasion of *Codium fragile* ssp. *tomentosoides* in northern Chile: a new threat for *Gracilaria* farming. Aquaculture 259: 202–210

Neushul P, Badash L (1998) Harvesting the Pacific: the blue revolution in China and the Philippines. Osiris 13:186–209

Neushul P, Wang Z (2000) Between the devil and the deep sea: C.K. Tseng, mariculture, and the politics of science in modern China. Isis 91:59–88

Norton B, Eames PC, Lo SNG (1998) Full-energy-chain analysis of greenhouse gas emissions for solar thermal electric power generation systems. Renew Energ 15:131–136

Nowak R (2008) Microalgae hold the key to the biofuel conundrum. New Sci 197(2641):12

Odada EO, Olago DO (2006) Challenges of an ecosystem approach to water monitoring and management of the African Great Lakes. Aquatic Ecosystem Health & Management 9:433–446
Odeh NA, Cockerill TT (2008) Life cycle GHG assessment of fossil fuel power plants with carbon capture and storage. Energ Policy 36:367–380
OECD-FAO (2007) OECD-FAO Agricultural Outlook 2007–2016. OECD, Paris
Ono E, Cuello JL (2006) Feasibility assessment of microalgal carbon dioxide sequestration technology with photobioreactor and solar collector. Biosyst Eng 95:597–606
Osowski S, Fahlenkamp H (2006) Regenerative energy production using energy crops. Ind Crop Prod 24:196–203
Patzek TW (2004) Thermodynamics of the corn-ethanol biofuel cycle. Crit Rev Plant Sci 23: 519–567
Patzek TW (2006) A first-law thermodynamic analysis of the corn-ethanol cycle. Nat Resour Res 15:255–270
Patzek TW, Pimentel D (2005) Thermodynamics of energy production from biomass. Crit Rev Plant Sci 24:327–364
Penner SS (2006) Steps toward the hydrogen economy. Energy 31:33–43
Pimentel D (2003) Ethanol fuels: energy balance, economics, and environmental impacts are negative. Nat Resour Res 12:127–134
Pimentel D, Herz M, Glickstein M, Zimmerman M, Allen R, Becker K, Evans J, Hussain B, Sarsfeld R, Grosfeld A, Seidel T (2002) Renewable energy: current and potential issues. BioScience 52:1111–1120
Prince RC, Kheshgi HS (2005) The photobiological production of hydrogen: potential efficiency and effectiveness as a renewable fuel. Crit Rev Microbiol 31:19–31
Ptasinski KJ, Hamelinck C, Kerkhof PJAM (2002) Exergy analysis of methanol from the sewage sludge process. Energ Convers Manage 43:1445–1457
Raines CA (2006) Transgenic approaches to manipulate the environmental responses of the $C_3$ carbon fixation cycle. Plant Cell Environ 29:331–339
Raoof B, Kaushik BD, Prasanna R (2006) Formulation of a low-cost medium for mass production of *Spirulina.* Biomass Bioenerg 30:537–542
Reijnders L (2006) Conditions for the sustainability of biomass based fuel use. Energ Policy 34:863–876
Reijnders L (2008) Do biofuels from microalgae beat biofuels from terrestrial plants? Trends Biotechnol 26:349–350
Reijnders L, Huijbregts MAJ (2003) Choices in calculating life cycle emissions of carbon containing gases associated with forest derived biofuels. J Clean Prod 11:527–532
Reijnders L, Huijbregts MAJ (2005) Life cycle emissions of greenhouse gases associated with burning animal wastes in countries of the European Union. J Clean Prod 13:51–56
Reijnders L, Huijbregts MAJ (2007) Life cycle greenhouse gas emissions, fossil fuel demand and solar energy conversion efficiency in European bioethanol production for automotive purposes. J Clean Prod 15:1806–1812
Reijnders L, Huijbregts MAJ (2008a) Palm oil and the emission of carbon-based greenhouse gases. J Clean Prod 16:477–482
Reijnders L, Huijbregts MAJ (2008b) Biogenic greenhouse gas emissions linked to the life cycles of biodiesel derived from European rapeseed and Brazilian soybeans. J Clean Prod 16: 1943–1948
Reinharz J (1985) Science in the service of politics: the case of Chaim Weizmann during the First World War. Engl Hist Rev 100:572–603
Renewable Fuels Agency (2008) The Gallagher review of the indirect effects of biofuels. Renewable Fuels Agency, St Leonards-on-Sea (East Sussex, UK)
Robèrt M, Hultén P, Frostell B (2007) Biofuels in the energy transition beyond peak oil. A macroscopic study of energy demand in the Stockholm transport system 2030. Energy 32:2089–2098
Romero E, Bautista J, Garcia-Martinez AM, Cremades O, Parrado J (2007) Bioconversion of corn distiller's dried grains with solubles (CDDGS) to extracellular proteases and peptones. Process Biochem 42:1492–1497

Ros J, Nagelhout D, Montfoort J (2009) New environmental policy for system innovation: Casus alternatives for fossil motor fuels. Appl Energ 86:243–250

Rosing MT, Bird DK, Sleep NH, Glassley W, Albarede F (2006) The rise of continents – an essay on the geologic consequences of photosynthesis. Palaeogeogr Palaeocl 232:99–113

Royal Society (2008) Sustainable biofuels: prospects and challenges. http://royalsociety.org

Rupprecht J, Hankamer B, Mussgnug JH, Ananyev G, Dismukes C, Kruse O (2006) Perspectives and advances of biological $H_2$ production in microorganisms. Appl Microbiol Biotechnol 72:442–449

Rydh CJ, Sandén BA (2005) Energy analysis of batteries in photovoltaic systems. Part II: energy return factors and overall battery efficiencies. Energ Convers Manage 46:1980–2000

Şahin F, Çakmakçi R, Kantar F (2004) Sugar beet and barley yields in relation to inoculation with $N_2$-fixing and phosphate solubilizing bacteria. Plant Soil 265:123–129

Samson R, Mani S, Boddey R, Sokhansanj S, Quesada D, Urquiaga S, Reis V, Lem CH (2005) The potential of $C_4$ perennial grasses for developing a global bioheat industry. Crit Rev Plant Sci 24:461–495

Savage DF, Way J, Silver PA (2008) Defossiling fuel: how synthetic biology can transform biofuel production. ACS Chem Biol 3(1):13–16

Sawayama S, Minowa T, Yokoyama S (1999) Possibility of renewable energy production and $CO_2$ mitigation by thermochemical liquefaction of microalgae. Biomass Bioenerg 17:33–39

Scragg AH, Illman AM, Carden A, Shales SW (2002) Growth of microalgae with increased calorific values in a tubular reactor. Biomass Bioenerg 23:67–73

Shimamatsu H (2004) Mass production of *Spirulina*, an edible microalga. Hydrobiologia 512: 39–44

Silveira AM, Victoria RL, Ballesker MV, De Camargo PB, Martinelli LA, Cassia Picollo M (2006) Simulation of the effects of land use change in soil carbon dynamics in the Piracicaba River basin. www.embrapa.br

Sims REH, Senelwa K, Maiava T, Bullock BT (1999) *Eucalyptus* species for biomass energy in New Zealand – Part II: coppice performance. Biomass Bioenerg 17:333–343

Sinclair TR, Muchow RC (1999) Radiation use efficiency. Adv Agron 65:215–265

Skjånes K, Lindblad P, Muller J (2007) Bio$CO_2$ – a multidisciplinary, biological approach using solar energy to capture $CO_2$ while producing $H_2$ and high value products. Biomol Eng 24: 405–413

Somerville C (2007) Biofuels. Curr Biol 17(4):R115–R119

Srinivasan S, Mosdale R, Stevens P, Yang C (1999) Fuel cells: reaching the era of clean and efficient power generation in the twenty-first century. Annu Rev Energ Env 24:281–328

Tang H, Qiu J, Van Ranst E, Li C (2006) Estimations of soil organic carbon storage in cropland of China based on DNDC model. Geoderma 134:200–206

Taylor-Pickard J (2008) The pros and cons of feeding DDGS. Feed Mix 16(3):28–29

Tilman D, Fargione J, Wolff B, D'Antonio C, Dobson A, Howarth R, Schindler D, Schlesinger WH, Simberloff D, Swackhamer D (2001) Forecasting agriculturally driven global environmental change. Science 292:281–284

Tripanagnostopoulos Y, Souliotis M, Battisti R, Corrado A (2006) Performance, cost and life-cycle assessment study of hybrid PVT/AIR solar systems. Prog Photovoltaics Res Appl 14:65–76

Troell M, Robertson-Andersson D, Anderson RJ, Bolton JJ, Maneveldt G, Halling C, Probyn T (2006) Abalone farming in South Africa: an overview with perspectives on kelp resources, abalone feed, potential for on-farm seaweed production and socio-economic importance. Aquaculture 257:266–281

Turner JA (2004) Sustainable hydrogen production. Science 305:972–974

Tyner WE (2008) The US ethanol and biofuels boom: its origins, current status, and future prospects. BioScience 58:646–653

Tzilivakis J, Warner DJ, May M, Lewis KA, Jaggard K (2005) An assessment of the energy inputs and greenhouse gas emissions in sugar beet (*Beta vulgaris*) production in the UK. Agr Syst 85:101–119

Ugwu CU, Aoyagi H, Uchiyama H (2008) Photobioreactors for mass cultivation of algae. Bioresour Technol 99:4021–4028

van den Broek R, Vleeshouwers L, Hoogwijk M, van Wijk A, Turkenburg W (2001) The energy crop growth model SILVA: description and application to eucalyptus plantations in Nicaragua. Biomass Bioenerg 21:335–349

Van der Meer JP (1983) The domestication of seaweeds. BioScience 33:172–176

van Eijck J, Romijn H (2008) Prospects for Jatropha biofuels in Tanzania: an analysis with strategic niche management. Energ Policy 36:311–325

Vasudevan PT, Briggs M (2008) Biodiesel production – current state of the art and challenges. J Ind Microbiol Biotechnol 35:421–430

Vedenov D, Wetzstein M (2008) Toward an optimal U.S. ethanol fuel subsidy. Energ Econ 30:2073–2090

von Blottnitz H, Curran MA (2007) A review of assessments conducted on bio-ethanol as a transportation fuel from a net energy, greenhouse gas, and environmental life cycle perspective. J Clean Prod 15:607–619

Vonshak A, Richmond A (1988) Mass production of the blue-green alga *Spirulina*: an overview. Biomass 15:233–247

Vunjak-Novakovic G, Kim Y, Wu X, Berzin I, Merchuk JC (2005) Air-lift bioreactors for algal growth on flue gas: mathematical modeling and pilot-plant studies. Ind Eng Chem Res 44:6154–6163

Wang B, Li Y, Wu N, Lan CQ (2008) $CO_2$ bio-mitigation using microalgae. Appl Microbiol Biotechnol 79:707–718

Weidema BP (1993) Market aspects in product life cycle inventory methodology. J Clean Prod 1:161–166

Wijffels RR (2008) Potential of sponges and microalgae for marine biotechnology. Trends Biotechnol 26:26–31

Wikfors GH, Ohno M (2001) Impact of algal research in aquaculture. J Phycol 37:968–974

Wilcox HA (1982) The ocean as a supplier of food and energy. Experientia 38:31–35

Wilhelm SW, Suttle CA (1999) Viruses and nutrient cycles in the sea. BioScience 49:781–788

Wright DG, Mullen RW, Thomason WE, Raun WR (2001) Estimated land area increase of agricultural ecosystems to sequester excess atmospheric carbon dioxide. Commun Soil Sci Plant 32:1803–1812

Xu H, Miao X, Wu Q (2006) High quality biodiesel production from a microalga *Chlorella protothecoides* by heterotrophic growth in fermenters. J Biotechnol 126:499–507

Yang YF, Feng CP, Inamori Y, Maekawa T (2004) Analysis of energy conversion characteristics in liquefaction of algae. Resour Conserv Recy 43:21–33

Yazdani SS, Gonzalez R (2007) Anaerobic fermentation of glycerol: a path to economic viability for the biofuels industry. Curr Opin Biotechnol 18:213–219

Yoon JH, Shin JH, Kim M, Sim SJ, Park TH (2006) Evaluation of conversion efficiency of light to hydrogen energy by *Anabaena variabilis*. Int J Hydrogen Energ 31:721–727

Zah R, Böni H, Gauch M, Hischier R, Lehman M, Wägner P (2007) Life cycle assessment of energy products: environmental impact assessment of biofuels. EMPA, St Gallen, Switzerland

Zhang Y, Habibi S, MacLean HL (2007) Environmental and economic evaluation of bioenergy in Ontario, Canada. J Air Waste Manag Assoc 57:919–933

Zwart K, Oudendag D, Ehlert P, Kuikman P (2006) Duurzaamheid en co-vergisting van dierlijke mest. Alterra, Wageningen, the Netherlands

# Chapter 3
# Non-energy Natural Resource Demand

## 3.1 Introduction

Several stocks of natural resources are highly important to biomass-for-energy, including transport biofuel, production. These are: soil and soil organic matter, nutrients and water. In this chapter, these will be discussed in turn as to their current status. We will also discuss the sustainable use of such resources. Sustainable is a term that has by now many meanings. Here, the term will be used in its original meaning in the modern environmental debate, linked to a steady state economy (Daly 1973; Gliessman 1989; Hueting and Reijnders 1998; Reijnders 2006). Thus, sustainable use of biomass is defined in this chapter as a type of use that can be continued indefinitely while maintaining the supply or availability of natural resources. Sustainability leads to limitations regarding the use of renewable resources which are characterized by large additions to the stock of these resources and concerning the use of resources that are geochemically scarce and formed in slow geological processes ('virtually non-renewable resources'). To allow for indefinite use, the usage of renewables should not exceed addition to stock, and resource quality should be maintained (Reijnders 2000). As to geochemically scarce virtually non-renewable resources, such as phosphate ore for which no substitution seems possible, the way to define sustainability is that wastes irretrievably lost should not substantially exceed the small addition to the stock by geological processes (Goodland and Daly 1996; Reijnders 2006). This requirement corresponds with a large reduction in current wastage.

## 3.2 Soil and Soil Organic Matter

Studies regarding annual crop production may provide clues about the factors necessary for a type of land-based biomass production that has high productivity and may be maintained indefinitely. Syers et al. (1997),Vance (2000) and Lal (2001a, b) have surveyed such studies and did show that one important factor in maintaining high productivity is soil conservation and the maintenance of high levels of organic matter in the upper layer of the soil. Loss of soil due to erosion ultimately leads to a strong decline in crop productivity (Lal 2001). Soil erosion is a major problem in annual crops, including the production of major current biofuel feedstocks such as corn, sugar cane and soybeans (Mahadevan 2008; Smeets et al. 2008), but also may be a problem in plantations and forests (Worrell and Hampson 1997; Perry 1998).

Soil organic matter is an important reserve for plant nutrients such as nitrogen (N) and phosphate (P). It improves soil structure and water-holding capacity (Kahle et al. 2002) and limits erosion (Troeh et al. 1999). Soil organic matter is also involved in weathering that extends the availability of nutrients (McBride 1994). Depletion of organic matter in soils ultimately results in a decrease in yields (Syers et al. 1997; Perry 1998). In many areas of the world, arable land currently shows a net loss of soil organic matter, if compared with its virgin or natural status (Cole et al. 1997; Ogle et al. 2005). Levels of soil organic carbon that are under long-term cultivation with annual crops tend to be lower than under native vegetation. On average, soil carbon levels under arable soils with a long history of cultivation tend to be approximately 18% lower under temperate dry conditions, approximately 30% lower under temperate moist or tropical dry conditions, and approximately 42% lower in tropical moist climates if compared with soils under native vegetation (Ogle et al. 2005). These reductions of carbon levels in soils have contributed to increased levels of greenhouse gases in the atmosphere. In many soils in tropical and subtropical areas, especially in Asia and sub-Saharan Africa, due to systematic excessive residue removal, soil organic carbon pools have decreased to levels that are conducive to soil degradation. Such soils also show substantially reduced production levels (Lal 2008).

Current agricultural practice often leads to further losses of soil carbon from arable soils. Net losses of soil carbon have been documented for the European Union (Vleeshouwers and Verhagen 2002), Eastern Canada (Gregorich et al. 2005), China (Li et al. 2003; Tang et al. 2006; Wright 2006), Nepal (Matthews and Pilbeam 2005; Shrestha et al. 2006), Brazil (Zinn et al. 2005; Jantalia et al. 2007), Sudan (Ardo and Olsson 2003), the southern Ethiopian highlands (Lemenih and Itanna 2004) and in West Africa (Ayanlaja et al. 1991; Ouattara et al. 2006; Bationo et al. 2007; Lufafa et al. 2008). From peaty arable soils losses may be especially high when there is deep drainage and intensive mechanical soil disturbance (Freibauer et al. 2004). Net carbon losses varying from 6 Mg in northern Norway up to 15 Mg C $ha^{-1}$ $year^{-1}$ in the tropics have been reported (Grønlund et al. 2006; Reijnders and Huijbregts 2007, 2008).

It has emerged that low crop yields, the absence of cover crops, low additions of crop residues, low additions of manure, mechanical tillage and high temperatures enhance the loss of carbon from arable soils (Vleeshouwers and Verhagen 2002; Pretty et al. 2002; Freibauer et al. 2004; Lal and Pimentel 2007; Valentin et al. 2008). There is evidence that excessive reliance of synthetic nitrogen fertilizers may be conducive to carbon loss (Kahn et al. 2007; Triberti et al. 2008). It has also been found that depending on the nature of the soil, climate, crops and crop rotation, crop residues management and tillage system, a partial (20–50%) removal of crop residues from the field reduces the pool of soil organic carbon, can exacerbate soil erosion hazard and negatively impact future yields of crops (Wilhelm et al. 2004; Lal 2005; Blanco-Canqui and Lal 2007; Lal 2008; Varvel et al. 2008). Models with parameters based on empirical data have been developed to estimate the impact of residue removal on soil organic carbon (e.g. Saffih-Hdadi and Mary 2008).

Sustainable use of soil and soil organic matter should be such that levels of soil organic matter do not decrease and that soil loss (erosion) should not exceed addition to topsoil stocks by natural processes. The latter may add between 0.004–0.5 mm of topsoil per year (Cannell and Hawes 1994). A variety of measures has been proposed to reduce erosion in annual cropping. These come under the umbrella of the term conservation tillage (Cannell and Hawes 1994; Lal 1997, 2001, 2008). They include the reduction of tillage (preferably to no-till or zero tillage), the use of cover crops and nitrogen-fixing legumes, intercropping, contour farming, increased return of harvest residues or residue mulches to the soil, use of manure, direct seeding, correcting effects of soil compaction due to vehicles and catching soil subject to water erosion on sloping soils by terracing or barriers (Gumbs 1993; Cannell and Hawes 1994; Lal 1997, 2001, 2008; Lal et al. 2007; Mills and Fey 2003). High inputs of carbon (residues, manure, compost) in soils that are subject to tillage are conducive to maintaining soil organic carbon levels (Jenkinson et al. 1990; Grace et al. 2006; Reijneveld et al. 2009).

In forests, limitations to harvesting and the use of heavy machinery on erodible soils and judicious planting may be necessary to prevent soil losses from exceeding additions to topsoil stocks (Pimentel et al. 1997b; Worrell and Hampson 1997). To maintain soil organic matter levels, in intensively managed forests and on plantations, intensive site preparation involving burning should be avoided, as this leads to volatilisation of soil carbon and prospective soil carbon (Perry 1998; Bauhus et al. 2002). Also, limitations to removal of harvest residues from forests may be necessary to maintain levels of soil organic matter (Worrell and Hampson 1997).

Soils to which crop residues are returned tend to store more soil organic carbon (and nitrogen) than plots where residues are taken away (Vance 2000; Mendham et al. 2002; Dolan et al. 2006; Epron et al. 2006). In this respect, there are two phenomena with opposite effects for $C_4$ and $C_3$ plants. Residues from plants that have $C_4$ photosynthesis seem less effective in contributing to soil carbon than the same amounts of residues from plants with $C_3$ photosynthesis (Wynn and Bird 2007). On the other hand, $C_4$ crops rather often generate relatively large amounts of below- and aboveground biomass, if compared with $C_3$ crops (Wright et al. 2001; Wilhelm et al. 2004). Vleeshouwers and Verhagen (2002) estimate that adding to arable soils the cereal straw that is currently taken away may, on average, increase European soil carbon levels by 0.15 Mg ha$^{-1}$ year$^{-1}$. Other studies have shown that full return of crop residues to arable soils in temperate climates may increase soil carbon levels by up to 0.7 Mg C ha$^{-1}$ year$^{-1}$ (Webb et al. 2003; Smith 2004; Freibauer et al. 2004; Rees et al. 2005). For maintaining soil carbon stocks in tropical soils, returning residues, application of other organic matter such as manure, shrub prunings and household composts, cover crops and fallows have been advocated (Lal and Bruce 1999; Bationo and Buerkert 2001; Nandwa 2001; Lufafa et al. 2008).

A changing climate will impact soil carbon stocks. In part, this impact is dependent on crop productivity. There is only limited empirical evidence about likely future crop productivity. Kim et al. (2007) have studied the $C_4$ crop corn and concluded that under elevated $CO_2$ concentration, productivity may remain unchanged. It has also been suggested that under elevated $CO_2$ concentrations in the atmosphere, productivity of $C_3$ plants may increase and that this may enhance soil carbon sequestration (Marhan et al. 2008), but increased temperature also leads to increased respiration in soils, and soil carbon dynamics may be impacted by changes in precipitation (Marhan et al. 2008). Overall effects may vary for different climate regions. In the European context, Vleeshouwers and Verhagen (2002) estimate that an increase in average temperature of 1 °C caused by an increase of $CO_2$ concentration may, ceteris paribus, lead to an average net loss of soil organic carbon of about $0.04\,\mathrm{Mg\,ha^{-1}\,year^{-1}}$.

In view of carbon losses, increased use of agricultural residues has been advocated in order to maintain (or restore) proper levels of soil organic carbon and ensure agroecosystem sustainability (Lal 1997; Duiker and Lal 1999; Lal 2001, 2008; Ouédraogo et al. 2006; Chivenge et al. 2007; Lal and Pimentel 2007). Adding lignocellulose to arable soil is more useful in this respect than more easily degradable carbon compounds such as sugars or starches (Sartori et al. 2006). Available evidence is limited but suggests that approximately 4–24% of carbon contained in residues of crops may be converted to refractory soil organic carbon in agricultural soils (Lal 1997; Follett et al. 2005; Razafimbelo et al. 2006; Triberti et al. 2008).

Crop residues contain cellulose in a matrix of lignin and hemicellulose. Lignin is, together with compounds such as cutins, suberins and tannins, largely responsible for humus formation in arable soils (Kirk 1971; Rasse et al. 2005) and in such soils is a major contributor to refractory soil organic carbon (Loveland and Webb 2003; Chapman and McCartney 2005). There is evidence that among the components of lignocellulose (lignin, hemicellulose and cellulose) in arable soils, lignin is by far the most refractory component (Melillo et al. 1989; Spaccini et al. 2000; Quénéa et al. 2006). Thus, lignin is more suitable for carbon sequestration in arable soils than hemicellulose. For this reason, removal of residues with high concentration of lignin (such as nut shells) may be expected to be more negative to arable soil carbon stocks than residues with a lower lignin level, such as wheat or rice straw.

Still, the presence of carbon compounds, which are more easily degraded than lignin (with a half life less than 1 year), in arable soil is also important for soil fertility and stability (Spaccini et al. 2000; Loveland and Webb 2003). The carbohydrates hemicellulose and cellulose in harvest residue belong to this category (Spaccini et al. 2000).

Against this background, systematic removal of all crop residues for biofuel production is not a good idea (Lal 2008; Reijnders 2008; Saffih-Hdadi and Mary 2008). Limitations on crop residue removal will have an upward effect on energy input into, and costs of residue collection for, biofuel production (Higgins et al. 2007). To the extent that crop residues are removed, there is furthermore a case for selecting crop residues for the production of the transport biofuel ethanol that have relatively high concentrations of hemicellulose and cellulose susceptible to conversion into ethanol. In the case of corn stover, this fraction consists of cobs, leaves and husks (Crofcheck and Montross 2004). The crop residue fraction that is relatively rich in lignin may be expected to be a relatively efficient contributor to refractory carbon in arable soils, but also contains a substantial amount of carbohydrates that are more easily degradable and contribute to soil fertility and stability. Thus, it may be that, for example, the scope of residue removal for ethanol production may be widened by selecting residues on the basis of their relative suitability for ethanol production and for the formation of refractory soil carbon, respectively.

Another option is to consider a return of processing 'wastes' of crop residues that are relatively rich in lignin. In generating ethanol from crop residues by enzymatic conversion (see Chap. 1), a residue emerges that is rich in lignin and also contains unreacted cellulose and hemicellulose (Mosier et al. 2005). It would seem worthwhile to consider applying this residue to arable soils. Such an application would serve the presence of refractory carbon in arable soil, while it may also contribute to the presence of more rapidly degradable carbohydrates. In doing so, one should limit or prevent undesirable side effects of adding this processing residue. A matter to consider in this respect is the accumulation potential of the residue for phenolic carbon compounds. Such accumulation may occur under anaerobic conditions, and this may have a negative effect on soil fertility (Olk et al. 2006). Ionic composition and pH of the processing residue are subject to limitation when use of the residue is to be sustainable (Mahmoudkhani et al. 2007). Also, one should be aware that lignin binds heavy metals such as cadmium much better than cellulose does (Basso et al. 2005). Thus, provisions should be in place to limit the flow of heavy metals to soils when the fraction that is rich in lignin is applied. If the processing residue has acceptable quality, it may well be that the amount of crop residue that can be removed from the field without a negative impact on soil characteristics can be increased. The quantitative and qualitative aspects of applying processing residues to arable soils would seem to merit further research. Finally, it should be noted that the refractory character of lignin in the arable soils studies cannot be generalized to all soils. There is, for instance, evidence that in lowland tree plantations in Costa Rica, litter decay increases with increasing lignin content (Raich et al. 2007).

## 3.3 Nutrients

Another factor important in productivity is the availability of sufficient mineral nutrients, such as fixed nitrogen (N), phosphorus (P), sulphur (S), potassium (K), calcium (Ca) and magnesium (Mg). As to the indefinite availability of sufficient nutrients, difficulties may well emerge (Manley and Richardson 1995; Sims and Riddell-Black 1998; Perry 1998; Ranger and Turpault 1999; Paré et al. 2002). These partly follow from the limitations of natural processes involved in making minerals available to the generation of biomass. These are deposition on soil and weathering (Hedin et al. 2003). On a time scale valid for forestry and cropping, there is only a very small addition to total reserves of minerals that can be made available to biomass due to geologic processes such as weathering (Ranger and Turpault 1999). Thus, the availability of nutrients from reserves generated by processes such as weathering may well go down over time on a time scale relevant to cropping and forestry due to losses linked to harvesting and erosion. External inputs of N and P in cropping and forestry are, moreover, dependent on the large-scale use of geochemically scarce natural resources (phosphate ore and fossil fuels) that are formed in slow geological processes, which does not allow for sustainable us. Fossil fuels are used to produce N-based synthetic fertilizers (Galloway et al. 2008). We will now first consider nutrients in forests and plantations and thereafter nutrients in arable soils.

### *3.3.1 Forests and Plantations*

Obtaining biofuels derived from plantations and forests heavily relies on including parts of trees, such as crowns, that have relatively high concentrations of nutrients (Manley and Richardson 1995; Perry 1998; Sims and Riddell-Black 1998; Paré et al. 2002; Rytter 2002). In the relatively young trees that characterize plantations, nutrient concentrations are, moreover, higher than in older trees (Rytter 2002). The use of feedstocks with high nutrient levels adds to losses of nutrients associated with common consequences of tree-harvesting practices such as erosion, increased leaching of nutrients and lowered rates of N-fixation by leguminous understory plants (Hamilton 1997; Heilman and Norby 1998; Richardson et al. 1999; Bernhardt et al. 2003).

Overall losses of nutrients may well have an impact on future productivity. For instance, studies of whole tree harvesting, with branches, tops and needles used as biofuels, as it is currently practiced in Sweden, show deficits in base cations (K, Mg and Ca) (Akselsson et al. 2007). Akselsson et al. (2007) suggest that compensatory fertilization with K, Mg and Ca is necessary to keep forestry sustainable. Studies in French forests show that the budget for Ca and probably Mg is negative, assuming a 60-year rotation time and a conservative scenario of biomass harvesting (Ranger and Turpault 1999). In the southern USA, P deficits have been noted (Pitman 2006). In Scandinavian forests, thinning involving whole tree removal has been found to cause significant reduction in stand volume increment linked with nutrient loss (Nord-Larsen 2002). And in tropical dry forests, repeated harvesting may well lead to reduced primary production due to a reduced capture of P from air (Lawrence et al. 2007). Keeping forest soil concentrations of nutrients in a steady state while removing feedstocks for biofuel production with relatively high concentrations of nutrients may, in the absence of nutrient amendments, force the application of long rotations or even an end to harvesting. In Sweden, harvesting trees from nutrient-poor soil is in fact discouraged (Manley and Richardson 1995).

There is also the option of the recycling of nutrients to soils. The extent to which this can be done depends on the use of biomass. For instance, fixed N is almost totally lost during combustion (Sander and Andrén 1997). On the other hand, it may be expected that fixed N can, to a large extent, be conserved in the production of methane and ethanol from biomass. In the context of ethanol production from switchgrass, Anex et al. (2007) have proposed a process that may recover about 78% of the fixed N input, which then may be recycled.

Nutrient elements other than N tend to be largely conserved in (fly and bottom) ash during burning and can be retrieved when proper controls are in place. In power plants, biomass is often co-fired with coal, and this will lead to ashes that are often considered unacceptable for nutrient application in forests or on arable soils (Reijnders 2005). When only biomass is burned, this may be different. It has been proposed to return ashes, especially for their base cations, to forest soils after burning forest-derived biomass (Hånell and Magnusson 2005; Pitman 2006; Ozolinèius et al. 2007). Pettersson et al. (2008) have suggested extracting phosphate from ashes for reuse as a nutrient. Also, digestate remaining after anaerobic conversion of biomass to methane, and fermentation residue remaining after converting lignocellulosic biomass into ethanol, may be returned to soils (e.g. Zwart et al. 2007; Reijnders 2006).

As yet, however, nutrient recycling is very limited. Ashes from burning biofuels are not usually returned to soils used for biofuel production, but largely diverted to other destinations, such as landfills (Reijnders 2005; Saikku et al. 2007). There may also be complications in returning nutrients. Most studies have focussed on the recycling of ash, and especially the recycling of base cations such as $Ca^{2+}$, $Mg^{2+}$ and $K^{+}$. Additions of such ashes to forests on mineral soils have shown disappointing effects, which have been linked to N deficiency (Augusto et al. 2008). It has been found that the chemistry of base cations in ashes tends to be different from the original chemistry in soils, and so are local concentrations of base cations in the soil after applications of ash. It has been argued that such differences may be limited by keeping temperatures between 600 and 900 °C during burning and using granulated ashes (Pitman 2006). Furthermore, it has been advised to apply ash to adult stands and not to seedlings (Augusto et al. 2008).

Experience with return of wood ash for its base cations to forest soil shows substantial side effects, for instance, on the levels of aluminium in soil solution and an increased soil emission of $CO_2$ (Maljanen et al. 2006; Ring et al. 2006), suppression of denitrification (Odlare and Pell 2009) and especially in the case of high fixed N presence in soils, increased leaching of nitrate (Pitman 2006). Reductions of Mn levels in biomass have been associated with wood ash recycling (Augusto et al. 2008). Moreover, hazardous elements, such as lead and cadmium, and hazardous organics, such as polycyclic aromatic hydrocarbons, may be present in ashes in substantial amounts. It has been found that even in the apparent absence of substantial anthropogenic contamination, levels of heavy metals in combustion ashes may be remarkably high (Reimann et al. 2008). For example, combustion ashes from South Norwegian birch and spruce wood ashes contained up to 1.3% lead and 203 $mg\,kg^{-1}$ cadmium (Reimann et al. 2008). Johansson and van Bavel (2003) and Enell et al.

(2008) looked at the presence of polycyclic aromatic hydrocarbons (PAHs) in wood ash and found that the concentration thereof in a substantial number of cases exceeded the standard applicable in Swedish forests of 2 $mg\,kg^{-1}$ of 16 PAHs.

Thus, high concentrations of inorganic and/or organic contaminants may represent a barrier to the sustainable recycling of nutrients. This may require input controls for burners (e.g. excluding wood with unacceptable levels of heavy metals), facilities for burning that minimize the formation of hazardous compounds such as polycyclic aromatic hydrocarbons and chlorinated dioxins and/or treatment of ashes to eliminate hazardous compounds. As pointed out in Sect. 3.2, in the case of wastes from lignocellulosic ethanol production, phenolic compounds, ionic composition and pH should be controlled, and the flow of heavy metals should be limited. All in all, it is unlikely that recycling of nutrients after biofuel processing and use can or will be as efficient as nutrient recycling in natural systems.

Regarding plantations, processes such as natural weathering and symbiotic N-fixation that are important in providing undisturbed forests with nutrients may well be less productive (Perry 1998). Intensive site preparation, common in plantations, involving burning biomass may negatively affect productivity as it leads to the volatilisation of nutrient N (Perry 1998). Because root systems in short rotation plantations may be less extensive than in undisturbed forests, the leakage of nutrients may well increase (Ong and Leakey 1999). Harvesting practices on plantations tend to increase denitrification and leaching, which may lead to increased nutrient deficits (Hamilton 1997; Heilman and Norby 1998). It is likely that erosion on plantations will exceed the value range 0.004–0.05 $Mg\,ha^{-1}\,year^{-1}$ that is found in undisturbed forests. Still, by judiciously planting and harvesting trees, it may be possible to keep erosion rates below the level of 1 $Mg\,ha^{-1}\,year^{-1}$ (Pimentel et al. 1997b).

Overall, as rates of biomass harvesting also tend to be much higher on plantations than in forests, large deficits in nutrients are to be expected. For instance, regarding aspen-for-fuel plantations, Rytter (2002) estimated the yearly deficit per hectare of N at 30 kg, for P at 4 kg, for Ca at 30 kg, for Mg at 4 kg and for S at 2.5 kg. Indeed, the productivity of growing short rotation trees strongly depends on external nutrient inputs (Adegbidi et al. 2001). Remarkably, current human activity, especially in industrialized countries, has led to increased environmental fluxes of wasted nutrients such as S, N and P, which reach soils via air and/or water (Smil 1991; Kvarnström and Nilsson 1999). For instance, in North America, unintended N depositions on soils may be up to 53 $hg\,ha^{-1}\,year^{-1}$, and in Europe up to 115 kg N $ha^{-1}\,year^{-1}$ (Heilman and Norby 1998). It is probable that the unintentional addition of nutrients to soils has been important to maintaining productivity in the absence of intentional nutrient amendments. For aspen plantations in Southern Sweden, it has, for instance, been calculated that deposition exceeds the yearly S deficit (2.5 $kg\,ha^{-1}$) and may cover a substantial part of the N deficit. In forests in Southern Sweden, there may still be accumulation of N with current biofuel harvesting practices (Akselsson et al. 2007). However, such unintended additions of nutrients to soils are not designed to fit all actual deficits in nutrients. For instance, though S deficits are compensated for in Nordic spruce forests, nutrient deficiencies for P, K and B still occur (Rytter 2002). Unintended additions of nutrients to soils may also cause new problems, especially when unintended nutrient additions are relatively high. They may well contribute to long-lasting eutrophication and acidification and deterioration of ground water quality (Galloway et al. 2008).

### 3.3.2 Agricultural Soils

Much agricultural production in Africa, Asia and South America is currently associated with a depletion of nutrients or 'nutrient mining' (de Koning et al. 1997; Syers et al. 1997; Sanchez 1999; Tilman et al. 2002). Increasing the use of crop residues for biofuel production may in practice exacerbate the latter problem (Troeh et al. 1999; Sauerbeck 2001). On the other hand, including woody perennials in agricultural strategies may well benefit both biomass-for-energy production and the long-term sustainability of food production, because there is evidence that woody perennials may recycle leached nutrients to near surface layers (Mele et al. 2003) and because several woody perennials are conducive to N-fixation (Sanchez 1999).

Long-term studies on the sustainability of crop production in industrialized countries (Vance 2000) are not easy to interpret as far as nutrients are concerned, because, as pointed out above, there are large unintended inputs of nutrients. However, such unintended inputs are usually well below those necessary for high-productivity cropping. Against this background, when harvest residues such as cereal and rape straw are used as biofuel, reuse of ash on arable land has been advocated (Sander and Andrén 1997).

Sustainable enhancement of biomass production can be achieved if there are ways to increase nutrient availability indefinitely (Vance 2000; Bhattacharya et al. 2003). For several nutrients, this is, technically speaking, not a major problem because the elements concerned are relatively abundant. Mg, K and Ca are in this category. However, P especially is geochemically scarce, and N nutrients are often generated by using geochemically scarce fossil fuels. This becomes even more of an issue, because increased biofuel production is expected to be partly based on increased yields of crops per hectare, which is linked to intensification of agriculture, including increased N and P inputs (Tilman et al. 2001; Searchinger et al. 2008).

Net natural inputs to farms of P, associated with weathering and deposition (Hedin et al. 2003), allow for very low primary productivity on farms (Newman 1997). Raising the availability of P if compared with pre-industrial times has mainly been achieved by relying on phosphate ore deposits, and there is no known alternative to that natural resource for doing so. So, sustainable use necessitates an extremely slow depletion of this stock. Ore deposits deplete rapidly when P used is not retrieved with high efficiency and fed back into biomass production. Non-retrieval associated with agriculture originates in harvesting, erosion (see Sect. 3.2) and leaching. Leaching is increased in agricultural land, if compared with soils under native forest (Williams and Melack 1997), and can be very high when soils are saturated with P (Liu et al. 2008). Preventing saturation leads to the need to restrict P additions to agricultural land.

Losses of P may also be linked with activities following harvesting. Much of the phosphate wastes associated with consumption and industrial processing of harvested biomass currently end up in wastewater. The recycling of such phosphate back into the economy is currently poorly developed (Sims and Riddell-Black 1998; Kvarnström and Nilsson 1999). One may also fail to retrieve P when there is burning of biomass. When biomass is burned, P ends up largely in ashes. When there is co-firing with, for instance, coal, such ashes are often considered unfit for agricultural use (Woodbury et al. 1999; Nugteren et al. 2001; Adriano et al. 2002; Reijnders 2005), whereas forced extraction of nutrients such as P from such ashes is not practiced. Even in the case of burning pure chicken manure, the composition of ashes may be such that nutrients are not fed back to agriculture (Reijnders and Huijbregts, 2005). Feed additives with high concentrations of trace elements such as Cu and Zn are responsible for this problem. On the other hand, in the case of biomass present in sewage sludge, a process has been developed for the fractionation and recovery of phosphate (Lundin et al. 2004), and this type of approach can, in principle, be applied to all wastes that contain substantial amounts of phosphate.

It has been estimated that the resulting net loss of P from the world's cropland is about $10.5 \times 10^6$ Mg per year, nearly one half of the amount of P that is extracted yearly as phosphate ore (Liu et al. 2008). Such large losses of P associated with production and use of biomass cannot be maintained indefinitely without jeopardizing adequate P levels in soils. If one will no longer be able to add substantial amounts of P to soils, primary productivity will ultimately plummet, negatively affecting both food and biofuel production (Newman 1997). So, indefinitely increased availability of P in soils is critically dependent on high-efficiency recycling of P involved in biomass production, while keeping soil concentrations of hazardous compounds below critical levels (Kvarnström and Nilsson 1999). A major effort is needed to apply this principle to biomass-for-energy.

The increased availability of nitrogen compounds to be used as fertilizer, if compared with the situation before the industrial revolution, is mainly based on the Haber synthesis, which converts fossil methane into ammonia (Galloway et al. 2008). As it stands, there is a large-scale leakage of added nitrogen compounds from biomass production systems such as plantations and annual crops. In well-managed intensive agriculture, the recovery of nitrogen in products is around 50% or less (Tinker 1997; Tilman et al. 2002). Moreover, N compounds present in biomass will largely get lost on burning. Basing the Haber synthesis on fossil fuels cannot be maintained indefinitely, as fossil carbon is virtually non-renewable. In this case, one may circumvent inputs of virtual non-renewables. For instance, hydrogen necessary for converting nitrogen present in air into N fertilizers can also be generated by hydrolysis powered by solar energy, a way of production that may be maintained indefinitely. There may also be scope for improved biogenic nitrogen fixation, which converts $N_2$ into plant nutrients and may partially replace fertilizer amendments.

## 3.4 Water

A last important non-energy natural resource for biomass production is water. The potential for biomass production on suitable soils is strongly influenced by freshwater availability (Tuskan 1998; Rabbinge and van Diepen 2000; Ryan et al. 2002; Kahle et al. 2002; Deckmyn et al. 2004), and in semi-arid areas, it may well be the main limiting factor (Ong and Leakey 1999). Of the yearly renewable stock of freshwater, approximately 30–50% is currently used in the economy (Postel et al. 1996; Raskin et al. 1997; Rockström 2003). Worldwide, growing crops is the main consumer of freshwater (Pimentel et al. 1997a). Even if growing of rain-fed crops is not considered to be use of water, growing crops accounts for at least 66% of current water use (Gleick 2000; Wallace 2002; van Dijk and Keenan 2007; Gordon et al. 2008; Zimmerman et al. 2008). Water consumption is projected to go up considerably, grossly paralleling growth of the world population and economic growth, much of the increase being associated with the expected increase in food and feed production (Gleick 2000; Berndes 2002; Swedish Environmental Advisory Council 2007).

Geographically speaking, water stocks are unevenly distributed. Currently, structural water shortages affect about 300–400 million people mainly in Africa and Asia in a band from China to North Africa (Gleick 2000; Wallace 2002). As much growth of the world population is expected in the same area, this does not bode well (Gleick 2000). But elsewhere there are problems, too. For instance, in the USA, roughly 20% of the irrigated area is supplied by groundwater pumped in excess of recharge (Tilman et al. 2002). Also, there are areas where climate change may lead to a structural decrease in rainfall, such as the south and east of Africa (Funk et al. 2008).

Large additional water requirements follow from expected population growth and changes in dietary habits, especially the increased consumption of animal produce (Rockström 2003; Falkenmark and Lannerstad 2005; Liu and Savenije 2008). The estimated size of the additional water requirement depends on the assumptions made. Assuming an intake of 3,000 kcal person$^{-1}$ day$^{-1}$ and using available predictions about world population growth, Rockström (2003) estimated an additional water requirement of 3,800 km$^3$ year$^{-1}$ in 2025 and of 5,800 km$^3$ year$^{-1}$ in 2050. Assuming business as usual, which does not include a substantial production of modern biomass-for-energy, it has been suggested that shortages of freshwater may well become a fact of life for up to 2.5–6.5 billion people by 2050 (World Water Council 2000; Wallace 2002). Expanding modern biomass-for-energy production may substantially exacerbate this trend.

The National Research Council of the USA has warned about more local water shortages due to the expanded production of corn for supplying ethanol (NRC 2007). This problem is especially pressing in the Western USA, where climate change threatens to exacerbate water shortages anyway (Barnett et al. 2008) and where there is already a substantial groundwater overdraft (Falkenmark and Lannerstad 2005). Excess water withdrawals may also be a problem in Brazilian sugar cane expansions (Dias de Oliveira et al. 2005), where hydrological flows are already much impacted by sugar cane cultivation (Gordon et al. 2008). India and China are expected to have increasing shortfalls in meeting national food demand due to water scarcity, necessitating increasing food imports (Falkenmark and Lannerstad 2005; Liu and Savenije 2008), and this leaves little scope for cropping feedstocks for transport biofuels.

In Hungary, there has been evidence that the use of cornstalks as biofuel worsened a drought by removing cover and reducing soil organic matter (Engelhaupt 2007). Berndes (2002) has pointed out that in the case of large-scale bioenergy production, a European country such as Poland may face absolute water scarcity. Competition between fast-growing *Eucalyptus* and food for water resources in Ethiopia has led to government restrictions on the former (Jagger and Pender 2003). Elsewhere, competition between short rotation trees and food crops for scarce water resources has also been noted (Ong and Leakey 1999; Sanchez 1999; Berndes 2002; Ong et al. 2002; van Dijk and Keenan 2007). On the other hand, when growing feedstocks are rain fed, there are some places subject to waterlogging and/or secondary salinization, where large uptakes of water by energy crops may be considered a benefit (Morris and Collopy 1999; Mahmood et al. 2001; Ryan et al. 2002; Palm et al. 2007), though it should be noted that the use of *Eucalyptus globulus* for this purpose in Western Australia has turned out to be disappointing (Sudmeyer and Simons 2008).

Data about seed-to-wheel water use are very patchy (Royal Society 2008). Data about water inputs into feedstock processing, especially, are largely lacking. Nevertheless, it is clear that biomass processing can lead to substantial water consumption. A case in point is the factories that produce ethanol from sugar cane in the relatively dry winter period in Brazil (Smeets et al. 2008). Also, in generating electricity from biomass in power plants, there is often water cooling that consumes large amounts of water (King and Webber 2008). However, it would seem likely that most of the seed-to-wheel water inputs concern the growing of feedstocks, in line with agriculture being the dominant consumer of water in the economy.

In view of increasing water scarcity, the water efficiency of transport biofuels requires considerable attention. As to water use in biomass processing, there is, for instance, scope for improving the water efficiency of Brazilian factories for the conversion of sugar cane into ethanol by increasing the water recycling rate and by replacing cane washing with dry cleaning of cane (Smeets et al. 2008). Similarly, consumption of cooling water by power plants processing biomass can be much reduced by, for example, switching to closed loop and air cooling systems (King and Webber 2008). The water efficiency in the growth of feedstocks is especially important (Hanegraaf et al. 1998; Rytter 2005; Almeida et al. 2007; Rockström et al. 2007). On average, the water efficiency of $C_4$ crop plants appears to be better than the water efficiency of $C_3$ crop plants (Cernusak et al. 2007; Heaton et al. 2008). Growing 1 kg of aboveground (dry) biomass of the $C_3$ crop wheat on average requires about 0.55 $m^3$ of water (calculated from data in Zwart and Bastiaanssen 2004), whereas 1 kg of dry biomass for the $C_4$ crops sugar cane or *Miscanthus* probably needs less than 0.2 $m^3$ (Rockström et al. 1999; Beale et al. 1999).

The water efficiency of soybeans and other annual oil crops is much poorer than the water efficiency of wheat (Liu and Savenije 2008). The microalga *Tetraselmis suecica*, which is considered for biofuel production, has been estimated to require 0.31–0.57 $m^3$ water $kg^{-1}$ dry weight biomass (Dismukes et al. 2008). As transport biofuels are the product of biomass processing, such biofuels will need a larger amount of water per kilogram of biofuel than per kilogram of biomass. In the case of ethanol based on cornstarch, while assuming an average water productivity of corn (Zwart and Bastiaanssen 2004) and a 0.294 kg ethanol yield from 1 kg of corn kernels (Kheshgi et al. 2000), for each kilogram of bioethanol, 1.9 $m^3$ of water is needed in cultivation, when all water use is allocated to ethanol production. In the case of wheat, the estimated water input in cultivation is about 3.1 $m^3$ water per kilogram of wheat-starch-based ethanol, when all water use is allocated to ethanol production. Water efficiency may change due to climate change linked to increasing greenhouse gas concentrations in the atmosphere. There is only very limited study thereof. A study on water use efficiency by a plantation producing willow as a feedstock for biofuels suggests that climate change may lead to a decrease in water efficiency (Tricker et al. 2009).

The difference in water efficiency between $C_3$ and $C_4$ crop plants is not the only noteworthy difference relevant to feedstock production. When the aim is to produce ethanol from carbohydrates, sugar crops tend, on average, to do so more efficiently than starch crops, whereas starchy root crops tend to be more water efficient than cereals (Rockström et al. 2007). In dry areas, the water efficiency of producing sugar by sugar beet may be better than that of sugar cane (Rytter 2005). Also, there may be 'drought-resistant' varieties of crops that would do better under such circumstances than other varieties (Rytter 2005). There is large variability in water efficiency between woody plants that may be used for biofuel production (Cernusak et al. 2007). Growing 1 kg of aboveground dry oil palm biomass requires approximately 1.5 $m^3$ of water (Rockström et al. 1999), and growing 1 kg of willow biomass (dry weight) requires 0.2–0.3 $m^3$ of water (Linderson et al. 2007). The olive tree is well adapted to drought stress (Sofo et al. 2008), whereas current varieties of *Eucalyptus* are less so (Dale and Dieters 2007). *Eucalyptus* biomass on rain-fed soil would require about 0.46 $m^3$ of water $kg^{-1}$ dry biomass (calculated from data in Stape et al. 2008 and Kheshgi et al. 2000). So, choice of feedstock may have a large impact on water consumption.

Soil water conservation is also dependent on tillage (and soil organic carbon levels) with conservation tillage conducive to water conservation (Wallace 2002; Rockström 2003). Mulching and intercropping to maximize canopy cover also help in increasing water productivity or crop yield for a specified input of water (Rockström 2003). So, for instance, on the Loess Plateau in China, no tillage with mulching does better in soil water conservation than conventional tillage, leading to substantially improved water use efficiency of winter wheat (Su et al. 2007). It has furthermore been shown that there are irrigation techniques that lead to higher water efficiencies than traditional furrow irrigation. These include variable application (Al-Kufaishi et al. 2006), subsoil irrigation and drip irrigation. Improvements in water efficiency may be considerable. Subsoil irrigation, with controlled diffusion of irrigation water from a clay pipe, has been shown to increase grain yield of Iranian winter wheat per kilogram of water by at least a factor of 2.5 (Banedjschafie et al. 2008). And in Uzbek cotton growing, which might supply cotton oil for biodiesel production, the improvement in yield by switching from furrow irrigation to drip irrigation was 35% to somewhat more than 100% (Ibragimov et al. 2007).

Apart from the quantitative impact on the availability of water, water quality may also be influenced by growing biofuels (NRC 2007). The expansion of corn production for the supply of ethanol is a case in point, as it leads to an increased nutrient and pesticide load of water resources (NRC 2007). And water discharges for algal biofuel production in open ponds may also be problematic in view of high pH, nutrient and/or salt levels present in pond water (Dismukes et al. 2008).

## 3.5 Sustainable Use of Natural Resources and Biomass Yields

Studies regarding the prospects for future modern biomass production tend to rely fully or overwhelmingly on land as the place where biomass is grown for this purpose. Studies with high estimates regarding the technical potential of biomass supply often have most of that potential met by energy crops that have high yearly yields per hectare (Hall and Rosillo-Calle 1998; Berndes et al. 2003; Hoogwijk et al. 2003; de Vries et al. 2007). Hoogwijk et al. (2003), for example, assumed a dry weight productivity of biomass-for-energy plantations on surplus agricultural land of up to 20 Mg $year^{-1}$ $ha^{-1}$. Such yields may be achieved. Actual experience shows that breeding efforts may increase biomass yields (dry weight) in the range of 6.7–11.3 Mg $ha^{-1}$ $year^{-1}$ to yields greater than 16 Mg (Volk et al. 2003). And there is much research aiming at further yield increases, for example, by lengthening the growing season without risking frost damage, limiting remobilisation of nutrients following senescence and improving drought resistance (Karp and Shield 2008). However, in general, highly productive species and varieties tend to be relatively inefficient in their resource use (Wood 1998), which is not in line with sustainable resource use (Pimentel et al. 2002; Reijnders 2006). Indeed, sustainable productivity is limited due to restrictions on water and nutrient use and the need to maintain adequate soil carbon levels.

In a first approximation to the levels of biomass production that may be produced in a sustainable way on land, it would seem useful to focus on natural net primary production (NNPP), which varies geographically (Havstad et al. 2007; Campbell et al. 2008). Kheshgi et al. (2000) estimated average natural NNPP on land at 4 Mg (= $10^6$ g) of dry biomass per year. Campbell et al. (2008), who studied abandoned agricultural soils, estimate that potential production rates on such soils average 4.3 Mg dry biomass $ha^{-1}$ $year^{-1}$. As pointed out earlier in this chapter, it may well be that recycling nutrients in the case of biofuel use is less efficient than in natural systems and that a part of carbon fixed in NNPP may be necessary to maintain soil organic carbon in a steady state. Thus, it is likely than on average, a lower amount of biomass can be harvested sustainably than 4–4.3 Mg $ha^{-1}$ $year^{-1}$. Pimentel et al. (2002) have suggested that in tropical and temperate areas, on average, approximately 3 Mg $ha^{-1}$ of woody biomass can be harvested in a sustainable way per year.

Again, there are geographical differences in sustainably harvestable biomass due to climate and water and nutrient availability (e.g. Nabuurs and Lioubimov 2000; Gough et al. 2008).

To get an idea of what a sustainable yield of feedstock may mean for energy supply, it would seem interesting to focus on agricultural land that has been abandoned (including currently fallow land). Field et al. (2008) and Campbell et al. (2008) estimate that the total area of such land is about $385–472 \times 10^6$ ha. We further assume that, after restoration of nutrients and soil organic matter, on these lands, a yearly sustainable yield of about 3 Mg (Pimentel et al. 2002) biomass with a lower heating value of $20\,\mathrm{MJ\,kg^{-1}}$ (Field et al. 2008) may be achieved. This would correspond with about 23–28 EJ ($= 10^{18}$ J) year$^{-1}$. As pointed out in Chap. 1, use of primary energy for the transport sector is currently about 100 EJ.

Another option that may be considered in the context of sustainable supply regards biofuels produced from what are currently 'wastes', such as organic urban wastes, biomass from forest remediation and residues from forestry and agriculture which are not used as animal feed. The worldwide amount of such wastes is currently estimated at between 50 and 100 EJ (Swedish Environmental Advisory Council 2007; Lysen and van Egmond 2008). Unfortunately, it is not clear how much thereof is necessary for maintaining the future productivity of arable lands and forests in line with the sustainability requirements for soil organic matter discussed in Sect. 3.2. However, even when only 10–20% thereof could be diverted to transport biofuel production, this would represent a substantial contribution to the transport fuel supply.

## References

Abrahamson LP, Robison DJ, Volk TA, White EH, Neuhauser EF, Benjamin WH, Peterson JM (1998) Sustainability and environmental issues associated with willow bioenergy development in New York (U.S.A.). Biomass Bioenerg 15:17–22

Adegbidi HG, Volk TA, White EH, Abrahamson LP, Briggs RD, Bickelhaupt DH (2001) Biomass and nutrient removal by willow clones in experimental bioenergy plantations in New York State. Biomass Bioenerg 20:399–411

Adriano DC, Weber J, Bolan NS (2002) Effects of high rates of coal fly ash on soil, turfgrass, and groundwater quality. Water Air Soil Poll 139:365–385

Akselsson C, Westling O, Sverdrup H, Gundersen P (2007) Nutrient and carbon budgets in forest soils as decision support in sustainable forest management. Forest Ecol Manag 238:167–174

Al-Kufaishi SA, Blackmore BS, Sourell H (2006) The feasibility of using variable rate water application under a central pivot irrigation system. Irrig Drain Syst 20:317–327

Almeida AC, Soares JV, Landsberg JJ, Rezende GD (2007) Growth and water balance of *Eucalyptus grandis* hybrid plantations in Brazil during a rotation for pulp production. Forest Ecol Manag 251:10–21

Alriksson A, Eriksson HM (2001) Distribution of Cd, Ck, Pb and Zn in soil and vegetation compartments in stands of five boreal tree species in N.E. Sweden. Water Air Soil Poll: Focus 1:461–475

Alvarez R (2005) Carbon stocks in Pampean soils: a simple regression model for estimation of carbon storage under nondegraded scenarios. Commun Soil Sci Plant 36:1583–1589

Anex RP, Lynd LR, Laser MS, Heggenstaller AH, Liebman M (2007) Potential for enhanced nutrient cycling through coupling of agricultural and bioenergy systems. Crop Sci 47:1327–1335
Ardo J, Olsson L (2003) Assessment of soil organic carbon in semi-arid Sudan using GIS and the CENTURY model. J Arid Environ 54:633–651
Augusto L, Bakker MR, Meredieu C (2008) Wood ash applications to temperate forest ecosystems – potential benefits and drawbacks. Plant Soil 306:181–198
Ayanlaja SA, Sanwo JO, Iwoye A (1991) Management of soil organic matter in the farming systems of the low land humid tropics of West Africa: a review. Soil Technol 4:265–279
Banedjschafie S, Bastani S, Widmoser P, Mengel K (2008) Improvement in water use and N fertilizer efficiency by subsoil irrigation of winter wheat. Eur J Agron 28:1–7
Barnett TP, Pierce DW, Hidalgo HG, Bonfils C, Santer BD, Das T et al. (2008) Human-induced changes in the hydrology of the western United States. Science 319:1080–1083
Basso MC, Cerella EG, Cukierman AL (2005) Cadmium uptake by lignocellulosic materials: effect of lignin content. Separ Sci Technol 39:1163–1175
Bationo A, Buerkert A (2001) Soil organic carbon management for sustainable land use in Sudano-Sahelian West Africa. Nutr Cycl Agroecosys 61:131–142
Bationo A, Kihara J, Vanlauwe B, Waswa B, Kimetu J (2007) Soil organic carbon dynamics, functions and management in West African agro-ecosystems. Agr Syst 94:13–25
Bauhus J, Khanna PK, Hopmans P, Weston C (2002) Is soil carbon a useful indicator of sustainable forest soil management? – a case study from native eucalypt forests of south-eastern Australia. Forest Ecol Manag 171:59–74
Beale CV, Morison JIL, Long SP (1999) Water use efficiency of $C_4$ perennial grasses in a temperate climate. Agr Forest Meteor 96:103–115
Berndes G (2002) Bioenergy and water – the implications of large-scale bioenergy production for water use and supply. Global Environ Chang 12:253–271
Berndes G, Hoogwijk M, van den Broek R (2003) The contribution of biomass in the future global energy supply: a review of 17 studies. Biomass Bioenerg 25:1–28
Bernesson S, Nilsson D, Hansson P (2006) A limited LCA comparing large- and small-scale production of ethanol for heavy engines under Swedish conditions. Biomass Bioenerg 30:46–57
Bernhardt ES, Likens GE, Buso DC, Driscoll CT (2003) In-stream uptake dampens effects of major forest disturbance on watershed nitrogen export. P Natl Acad Sci USA 100:10304–10308
Berthiaume R, Bouchard C, Rosen MA (2001) Exergetic evaluation of the renewability of a biofuel. Exergy 1:256–268
Bewket W, Stroosnijder L (2003) Effects of agroecological land use succession on soil properties in Chemoga watershed, Blue Nile basin, Ethiopia. Geoderma 111:85–98
Bhattacharya SC, Abdul Salam P, Pham HL, Ravindranath NH (2003) Sustainable biomass production for energy in selected Asian countries. Biomass Bioenerg 25:471–482
Blanco-Canqui H, Lal R (2007) Soil and crop response to harvesting corn residues for biofuel production. Geoderma 141:355–362
Buckley JC, Schwarz PM (2003) Renewable energy from gasification of manure: an innovative technology in search of fertile policy. Environ Monit Assessment 84:111–127
Butterworth B (2006) Biofuels from waste: closed loop sustainable fuel production. Refocus 7(3):60–61
Campbell JE, Lobell DB, Genova RC, Field CB (2008) The global potential of bioenergy on abandoned agriculture lands. Environ Sci Technol 42:5791–5794
Cannell RQ, Hawes JD (1994) Trends in tillage practices in relation to sustainable crop production with special reference to temperate climates. Soil Till Res 30:245–282
Cernusak LA, Aranda J, Marshall JD, Winter K (2007) Large variation in whole-plant water-use efficiency among tropical tree species. New Phytol 173:294–305
Chapman S, McCartney D (2005) Composting residuals from a strawboard manufacturing facility. Compost Sci Util 13:90–97
Chivenge PP, Murwira HK, Giller KE, Mapfumo P, Six J (2007) Long-term impact of reduced tillage and residue management on soil carbon stabilization: implications for conservation agriculture on contrasting soils. Soil Till Res 94:328–337

Chum HL, Overend RP (2001) Biomass and renewable fuels. Fuel Process Technol 71:187–195

Cole CV, Duxbury J, Freney J, Heinemeyer O, Minami K, Mosier A, Paustian K, Rosenberg N, Sampson N, Sauerbeck D, Zhao Q (1997) Global estimates of potential mitigation of greenhouse gas emissions by agriculture. Nutr Cycl Agroecosys 49:221–228

Crofcheck CL, Montross MD (2004) Effect of stover fraction on glucose production using enzymatic hydrolysis. T Am Soc Agr Eng 47:841–844

Dale G, Dieters M (2007) Economic returns from environmental problems: breeding salt- and drought-tolerant eucalypts for salinity abatement and commercial forestry. Ecol Eng 31: 175–182

Daly HE (1973) Toward a steady-state economy. Freeman, San Francisco

Deckmyn G, Laureysens I, Garcia J, Muys B, Ceulemans R (2004) Poplar growth and yield in short rotation coppice: model simulations using the process model SECRETS. Biomass Bioenerg 26:221–227

de Koning GHJ, van de Kop PJ, Fresco LO (1997) Estimates of sub-national nutrient balances as sustainability indicators for agro-ecosystems in Ecuador. Agr Ecosyst Environ 65:127–139

Demirbaş A (2007) Progress and recent trends in biofuels. Prog Energ Combust 33:1–18

de Vries BJM, van Vuuren DP, Hoogwijk MM (2007) Renewable energy sources: their global potential for the first-half of the 21st century at a global level: an integrated approach. Energ Policy 35:2590–2610

Dias de Oliveira M, Vaughan BE, Rykiel EJ Jr (2005) Ethanol as fuel: energy, carbon dioxide balances, and ecological footprint. BioScience 55:593–602

Dismukes GC, Carrieri D, Bennette N, Ananyev GM, Posewitz MC (2008) Aquatic phototrophs: efficient alternatives to land-based crops for biofuels. Curr Opin Biotechnol 19:235–240

Dolan MS, Clapp CE, Allmaras RR, Baker JM, Molina JAE (2006) Soil organic carbon and nitrogen in a Minnesota soil as related to tillage, residue and nitrogen management. Soil Till Res 89:221–231

Droppelmann KJ, Lehmann J, Ephrath JE, Berliner PR (2000) Water use efficiency and uptake patterns in a runoff agroforestry system in an arid environment. Agroforest Syst 49:223–243

Duiker SW, Lal R (1999) Crop residue and tillage effects on carbon sequestration in a Luvisol in central Ohio. Soil Till Res 52:73–81

Enell A, Fuhrman F, Lundin L, Warfvinge P, Thelin G (2008) Polycyclic aromatic hydrocarbons in ash: determination of total and leachable concentrations. Environ Pollut 152:285–292

Engelhaupt E (2007) Biofueling water problems. Environ Sci Technol 15:7593–7595

Epron D, Nouvellon Y, Deleporte P, Ifo S, Kazotti G, M'Bou AT, Mouvondy W, Saint Andre L, Roupsard O, Jourdan C, Hamel O (2006) Soil carbon balance in a clonal Eucalyptus plantation in Congo: effects of logging on carbon inputs and soil $CO_2$ efflux. Glob Change Biol 12: 1021–1031

European Commission (2006) Commission urges new drive to boost production of biofuels. Press release reference IP/06/135

Falkenmark M, Lannerstad M (2005) Consumptive water use to feed humanity – curing a blind spot. Hydrol Earth Syst Sci 9:15–28

Felker P, Guevara JC (2003) Potential of commercial hardwood forestry plantations in arid lands – an economic analysis of *Prosopis* lumber production in Argentina and the United States. Forest Ecol Manag 186:271–286

Field CB, Campbell JE, Lobell DB (2008) Biomass energy: the scale of the potential resource. Trends Ecol Evol 23:65–72

Fischer G, Schrattenholzer L (2001) Global bioenergy potentials through 2050. Biomass Bioenerg 20:151–159

Fleming JS, Habibi S, MacLean H (2006) Investigating the sustainability of lignocellulose-derived fuels for light-duty vehicles. Transport Res D-Tr E 11:146–159

Follett RF, Castellanos JZ, Buenger ED (2005) Carbon dynamics and sequestration in an irrigated vertisol in Central Mexico. Soil Till Res 83:148–158

Freibauer A, Rounsevell MDA, Smith P, Verhagen J (2004) Carbon sequestration in the agricultural soils of Europe. Geoderma 122:1–23

Fung PYH, Kirschbaum MUF, Raison RJ, Stucley C (2002) The potential for bioenergy production from Australian forests, its contribution to national greenhouse targets and recent developments in conversion processes. Biomass Bioenerg 22:223–236

Funk C, Dettinger MD, Michaelsen JC, Verdin JP, Brown ME, Barlow M, Hoell A (2008) Warming of the Indian Ocean threatens eastern and southern African food security but could be mitigated by agricultural development. P Natl Acad Sci USA 105:11081–11086

Galloway JN, Townsend AR, Erisman JW, Bekunda M, Cai Z, Freney JR, Martinelli LA, Seitzinger SP, Sutton MA (2008) Transformation of the nitrogen cycle: recent trends, questions, and potential solutions. Science 320:889–892

Giampietro M, Ulgiati S, Pimentel D (1997) Feasibility of large-scale biofuel production: does an enlargement of scale change the picture? BioScience 47:587–600

Gleick PH (2000) The world's water 2000–2001: the biennial report on freshwater resources. Island Press, Washington, DC

Gliessman SR (1989) Quantifying the agroecological component of sustainable agriculture: a goal. In: Gliessman SR (ed) Agroecology – researching the ecological basis for sustainable agriculture. Springer Verlag, New York, pp 366–370

Goldemberg J, Teixeira Coelho S (2004) Renewable energy – traditional biomass vs. modern biomass. Energ Policy 32:711–714

Goldemberg J, Teixeira Coelho S, Guardabassi P (2008) The sustainability of ethanol production from sugarcane. Energ Policy 36:2086–2097

Goodland R, Daly H (1996) Environmental sustainability: universal and non-negotiable. Ecol Appl 6:1002–1017

Gordon LJ, Peterson GD, Bennett EM (2008) Agricultural modifications of hydrological flows create ecological surprises. Trends Ecol Evol 23:211–219

Gough CM, Vogel CS, Schmid HP, Curtis PS (2008) Controls on annual forest carbon storage: lessons from the past and predictions for the future. BioScience 58:609–622

Grace PR, Colunga-Garcia M, Gage SH, Robertson GP, Safir GR (2006) The potential impact of agricultural management and climate change on soil organic carbon of the north central region of the United States. Ecosystems 9:816–827

Greer D (2005) Spinning straw into fuel. BioCycle 46:61–67

Gregorich EG, Rochette P, VandenBygaart AJ, Angers DA (2005) Greenhouse gas contributions of agricultural soils and potential mitigation practices in Eastern Canada. Soil Till Res 83:53–72

Grønlund A, Sveistrup TE, Søvik AK, Rasse DP, Kløve B (2006) Degradation of cultivated peat soils in northern Norway based on field scale $CO_2$, $N_2O$ and $CH_4$ emission measurements. Arch Agron Soil Sci 52:149–159

Gumbs FA (1993) Tillage methods and soil and water conservation in the Caribbean. Soil Till Res 27:341–354

Hall DO, Rosillo-Calle F (1998) Biomass resources other than wood. World Energy Council, London

Hall DO, Rosillo-Calle F, Williams RH, Woods J (1993) Biomass for energy: supply prospects. In: Johansson TB, Kelly H, Reddy AKN, Williams RH (eds) Renewable energy sources for fuels and electricity. Island Press, Washington, DC, pp 593–651

Hamilton C (1997) The sustainability of logging in Indonesia's tropical forests: a dynamic input-output analysis. Ecol Econ 21:183–195

Hanegraaf MC, Biewinga EE, van der Bijl G (1998) Assessing the ecological and economic sustainability of energy crops. Biomass Bioenerg 15:345–355

Hånell B, Magnusson T (2005) An evaluation of land suitability for forest fertilization with biofuel ash on organic soils in Sweden. Forest Ecol Manage 209:43–55

Havstad KM, Peters DPC, Skaggs R, Brown J, Bestelmeyer B, Fredrickson E, Herrick J, Wright J (2007) Ecological services to and from rangelands of the United States. Ecol Econ 64:261–268

Heaton EA, Flavell RB, Mascia PN, Thomas SR, Dohleman FG, Long SP (2008) Herbaceous energy crop development: recent progress and future prospects. Curr Opin Biotechnol 19: 202–209

Hedin LO, Vitousek PM, Matson PA (2003) Nutrient losses over four million years of tropical forest development. Ecology 84:2231–2255

Heilman P, Norby RJ (1998) Nutrient cycling and fertility management in temperate short rotation forest systems. Biomass Bioenerg 14:361–370

Heller MC, Keoleian GA, Volk TA (2003) Life cycle assessment of a willow bioenergy cropping system. Biomass Bioenerg 25: 147–165

Higgins D, Kok H, Kruger C (2007) Bioenergy and soil carbon sequestration. BioCycle 48:23–24

Hoogwijk M, Faaij A, van den Broek R, Berndes G, Gielen D, Turkenburg W (2003) Exploration of the ranges of the global potential of biomass for energy. Biomass Bioenerg 25:119–133

Hueting R, Reijnders L (1998) Sustainability is an objective concept. Ecol Econ 27:139–147

Ibragimov N, Evett SR, Esanbekov Y, Kamilov BS, Mirzaev L, Lamers JPA (2007) Water use efficiency of irrigated cotton in Uzbekistan under drip and furrow irrigation. Agr Water Manage 90:112–120

Jackson RB, Jobbágy EG, Avissar R, Roy SB, Barrett DJ, Cook CW, Farley KA, le Maitre DC, McCarl BA, Murray BC (2005) Trading water for carbon with biological carbon sequestration. Science 310:1944–1947

Jagger P, Pender J (2003) The role of trees for sustainable management of less-favored lands: the case of eucalyptus in Ethiopia. Forest Policy Econ 5:83–95

Jantalia CP, Resck DVS, Alves BJR, Zotarelli L, Urquiaga S, Boddey RM (2007) Tillage effect on C stocks of a clayey oxisol under a soybean-based crop rotation in the Brazilian Cerrado region. Soil Till Res 95:97–109

Jenkinson DS, Andrew SPS, Lynch JM, Goss MJ, Tinker PB (1990) The turnover of organic carbon and nitrogen in soil. Philos T R Soc B 329:361–368

Johansson I, van Bavel B (2003) Levels and patterns of polycyclic aromatic hydrocarbons in incineration ashes. Sci Total Environ 311:221–231

Johnson JMF, Reicosky DC, Allmaras RR, Sauer TJ, Venterea RT, Dell CJ (2005) Greenhouse gas contributions and mitigation potential of agriculture in the central USA. Soil Till Res 83:73–94

Kabel MA, Bos G, Zeevalking J, Voragen AGJ, Schols HA (2007) Effect of pretreatment severity on xylan solubility and enzymatic breakdown of the remaining cellulose from wheat straw. Bioresour Technol 98:2034–2042

Kahle P, Belau L, Boelcke B (2002) Auswirkungen eines 10-jährigen Miscantusanbaus auf ausgewählte Eigenschaften eines Mineralbodens in Nordostdeutschland. J Agron Crop Sci 188: 43–50

Kahn SA, Mulvaney RL, Ellsworth TR, Boast CW (2007) The myth of nitrogen fertilization for soil carbon sequestration. J Environ Qual 36:1821–1832

Kaltschmitt M, Reinhardt GA, Stelzer T (1997) Life cycle analysis of biofuels under different environmental aspects. Biomass Bioenerg 12:121–134

Karp A, Shield I (2008) Bioenergy from plants and the sustainable yield challenge. New Phytol 179:15–32

Kheshgi HS, Prince RC, Marland G (2000) The potential of biomass fuels in the context of global climate change: focus on transportation fuels. Annu Rev Energ Env 25:199–244

Kim S, Gitz DC, Sicher RC, Baker JT, Timlin DJ, Reddy VR (2007) Temperature dependence of growth, development, and photosynthesis in maize under elevated $CO_2$. Environ Exp Bot 61:224–236

King CW, Webber ME (2008) The water intensity of the plugged-in automotive economy. Environ Sci Technol 42:4305–4311

Kirk TK (1971) Effects of microorganisms on lignin. Annu Rev Microbiol 9:185–210

Körbitz W (2002) New trends in developing biodiesel world-wide. Austrian Biofuels Institute, Vienna

Kraxner F, Nilsson S, Obersteiner M (2003) Negative emissions from bioenergy use, carbon capture and sequestration (BECS) – the case of biomass production by sustainable forest management from semi-natural temperate forests. Biomass Bioener 24:285–296

Krotscheck C, König F, Obernberger I (2000) Ecological assessment of integrated bioenergy systems using the Sustainable Process Index. Biomass Bioenerg 18:341–368

Kvarnström E, Nilsson M (1999) Reusing phosphorus: engineering possibilities and economic realities. J Econ Issues 33:393–402

Lal R (1997) Residue management, conservation tillage and soil restoration for mitigating greenhouse effect by $CO_2$ enrichment. Soil Till Res 43:81–107

Lal R (2001a) World cropland soils as a source or sink for atmospheric carbon. Adv Agron 71: 145–191

Lal R (2001b) Managing world soils for food security and environmental quality. Adv Agron 74:155–192

Lal R (2005) World crop residues production and implications of its use as a biofuel. Environ Int 31:575–584

Lal R (2008) Crop residues as soil amendments and feedstock for bioethanol production. Waste Manage 28:747–758

Lal R, Bruce JP (1999) The potential of world cropland soils to sequester C and mitigate the greenhouse effect. Environ Sci Policy 2:177–185

Lal R, Pimentel D (2007) Biofuels from crop residues. Soil Till Res 93:237–238

Lal R, Reicosky DC, Hanson JD (2007) Evolution of the plow over 10,000 years and the rationale for no-till farming. Soil Till Res 93:1–12

Lawrence D, D'Odorico P, Diekmann L, DeLonge M, Das R, Eaton J (2007) Ecological feedbacks following deforestation create the potential for a catastrophic ecosystem shift in tropical dry forest. P Natl Acad Sci USA 104:20696–20701

Lemenih M, Itanna F (2004) Soil carbon stocks and turnovers in various vegetation types and arable lands along an elevation gradient in southern Ethiopia. Geoderma 123:177–188

Li C, Zhuang Y, Frolking S, Galloway J, Harriss R, Moore B, Schimel D, Wang X (2003) Modeling soil organic carbon change in croplands of China. Ecol Appl 13:327–336

Linderson M, Iritz Z, Lindroth A (2007) The effect of water availability on stand-level productivity, transpiration, water use efficiency and radiation use efficiency of field-grown willow clones. Biomass Bioenerg 31:460–468

Liu J, Savenije HHG (2008) Food consumption patterns and their effect on water requirement in China. Hydrol Earth Syst Sci 12:887–898

Liu Y, Villalba G, Ayres RU, Schröder H (2008) Global phosphorus flows and environmental impacts from a consumption perspective. J Ind Ecol 12:229–247

Loveland P, Webb J (2003) Is there a critical level of organic matter in the agricultural soils of temperate regions: a review. Soil Till Res 70:1–18

Lufafa A, Bolte J, Wright D, Khouma M, Diedhiou I, Dick RP, Kizito F, Dossa E, Noller JS (2008) Regional carbon stocks and dynamics in native woody shrub communities of Senegal's peanut basin. Agr Ecosyst Environ 128:1–11

Lundin M, Olofsson M, Pettersson GJ, Zetterlund H (2004) Environmental and economic assessment of sewage sludge handling options. Resour Conserv Recy 41:255–278

Lynd LR, Wyman CE, Gerngross TU (1999) Biocommodity Engineering. Biotechnol Prog 15: 777–793

Lysen E, van Egmond S (eds) (2008) Assessment of biomass potentials and their links to food, water, biodiversity, energy demand and economy. http://www.mnp.nl/en/publications/2008/Assessment_of_global_biomass_potentials_MainReport.html

MacLean HL, Lave LB (2003) Evaluating automobile fuel/propulsion system technologies. Prog Energ Combust 29:1–69

Mahadevan R (2008) The high price of sweetness: the twin challenges of efficiency and soil erosion in Fiji's sugar industry. Ecol Econ 66:468–477

Mahmood K, Morris J, Collopy J, Slavich P (2001) Groundwater uptake and sustainability of farm plantations on saline sites in Punjab province, Pakistan. Agr Water Manag 48:1–20

Mahmoudkhani M, Richards T, Theliander H (2007) Sustainable use of biofuel by recycling ash to forests: treatment of biofuel ash. Environ Sci Technol 41:4118–4123

Maljanen M, Nykänen H, Moilanen M, Martikainen PJ (2006) Greenhouse gas fluxes of coniferous forest floors as affected by wood ash addition. Forest Ecol Manage 237:143–149

Manley A, Richardson J (1995) Silviculture and economic benefits of producing wood energy from conventional forestry systems and measures to mitigate negative impacts. Biomass Bioenerg 9:89–105

Marhan S, Demin D, Erbs M, Kuzyakov Y, Fangmeier A, Kandeler E (2008) Soil organic matter mineralization and residue decomposition of spring wheat grown under elevated $CO_2$ atmosphere. Agr Ecosyst Environ 123:63–68

Matthews RB, Pilbeam C (2005) Modelling the long-term productivity and soil fertility of maize/millet cropping systems in the mid-hills of Nepal. Agr Ecosyst Environ 111:119–139

McBride MB (1994) Environmental chemistry of soils. Oxford University, Oxford

Mele PM, Yunusa IAM, Kingston KB, Rab MA (2003) Response of soil fertility indices to a short phase of Australian woody species: continuous annual crop rotations or a permanent pasture. Soil Till Res 72:21–30

Melillo JM, Aber JD, Linkins AE, Ricca A, Fry B, Nadelhoffer KJ (1989) Carbon and nitrogen dynamics along the decay continuum: plant litter to soil organic matter. Plant Soil 115:189–198

Mendham DS, Sankaran KV, O'Connell AM, Grove TS (2002) *Eucalyptus globulus* harvest residue management effects on soil carbon and microbial biomass at 1 and 5 years after plantation establishment. Soil Biol Biochem 34:1903–1912

Mills AJ, Fey MV (2003) Declining soil quality in South Africa: effects of land use on soil organic matter and surface crusting. S Afr J Sci 99:429–436

Moreira JR (2006) Global biomass energy potential. Mitigation Adaptation Strategies Global Change 11:313–333

Morris JD, Collopy JJ (1999) Water use and salt accumulation by *Eucalyptus camaldulensis* and *Casuarina cunninghamiana* on a site with shallow saline groundwater. Agr Water Manag 39:205–227

Mosier N, Wyman C, Dale B, Elander R, Lee YY, Holtzapple M, Ladisch M (2005) Features of promising technologies for pretreatment of lignocellulosic biomass. Bioresour Technol 96: 673–686

Nabuurs GJ, Lioubimov AV (2000) Future development of the Leningrad region forests under nature-oriented forest management. Forest Ecol Manag 130:235–251

Nandwa SM (2001) Soil organic carbon (SOC) management for sustainable productivity of cropping and agro-forestry systems in Eastern and Southern Africa. Nutr Cycl Agroecosyst 61: 143–158

Newman EI (1997) Phosphorus balance of contrasting farming systems, past and present. Can food production be sustainable? J Appl Ecol 34:1334–1347

Nord-Larsen T (2002) Stand and site productivity response following whole-tree harvesting in early thinnings of Norway spruce. Biomass Bioenerg 23:1–12

NRC (National Research Council) (2007) Water implications of biofuels production in the United States. National Academy Press, Washington, DC

Nugteren HW, Janssen-Jurkovicova M, Scarlett B (2001) Improvement of environmental quality of coal fly ash by applying forced leaching. Fuel 80:873–877

Odlare M, Pell M (2009) Effect of wood fly ash and compost on nitrification and denitrification in agricultural soil. Appl Energ 86:74–80

Ogle SM, Breidt FJ, Paustian K (2005) Agricultural management impacts on soil organic carbon storage under moist and dry climatic conditions of temperate and tropical regions. Biogeochemistry 72:87–121

Olk DC, Cassman KG, Schmidt-Rohr K, Anders MM, Mao J, Deenik JL (2006) Chemical stabilization of soil organic nitrogen by phenolic lignin residues in anaerobic agroecosystems. Soil Biol Biochem 38:3303–3312

Ong CK, Leakey RRB (1999) Why tree-crop interactions in agroforestry appear at odds with tree-grass interactions in tropical savannahs. Agroforest Syst 45:109–129

Ong CK, Wilson J, Deans JD, Mulayta J, Raussen T, Waja-Musukwe N (2002) Tree-crop interactions: manipulation of water use and root function. Agr Water Manag 53:171–186

Ouattara B, Ouattara K, Serpantié G, Mando A, Sédogo MP, Bationo A (2006) Intensity cultivation induced effects on soil organic carbon dynamic in the western cotton area of Burkina Faso. Nutr Cycl Agroecosys 76:331–339

Ouédraogo E, Mando A, Stroosnijder L (2006) Effects of tillage, organic resources and nitrogen fertiliser on soil carbon dynamics and crop nitrogen uptake in semi-arid West Africa. Soil Till Res 91:57–67

Ozkan B, Akcaoz H, Fert C (2004) Energy input-output analysis in Turkish agriculture. Renew Energ 29:39–51

Ozolinèius R, Varnagiryté-Kabašinskiene I, Stakènas V, Mikšys V (2007) Effects of wood ash and nitrogen fertilization on Scots pine crown biomass. Biomass Bioenerg 31:700–709

Palm C, Sanchez P, Ahamed S, Awiti A (2007) Soils: a contemporary perspective. Annu Rev Environ Resour 32:99–129

Paré D, Rochon P, Brais S (2002) Assessing the geochemical balance of managed boreal forests. Ecol Indic 1:293–311

Patzek TW, Pimentel D (2005) Thermodynamics of energy production from biomass. Crit Rev Plant Sci 24:327–364

Perry DA (1998) The scientific basis of forestry. Annu Rev Ecol Syst 29:435–466

Pettersson A, Åmand L, Steenari B (2008) Leaching of ashes from co-combustion of sewage sludge and wood – Part I: recovery of phosphorus. Biomass Bioenerg 32:224–235

Pimentel D (2003) Ethanol fuels: energy balance, economics, and environmental impacts are negative. Nat Resour Res 12:127–134

Pimentel D, Moran MA, Fast S, Weber G, Bukantis R, Balliett L, Boveng P, Cleveland C, Hindman S, Young M (1981) Biomass energy from crop and forest residues. Science 212:1110–1115

Pimentel D, Houser J, Preiss E, White O, Fang H, Mesnick L, Barsky T, Tariche S, Schreck J, Alpert S (1997a) Water resources: agriculture, the environment, and society. An assessment of the status of water resources. BioScience 47:97–106

Pimentel D, McNair M, Buck L, Pimentel M, Kamil J (1997b) The value of forests to world food security. Hum Ecol 25:91–120

Pimentel D, Herz M, Glickstein M, Zimmerman M, Allen R, Becker K, Evans J, Hussain B, Sarsfeld R, Grosfeld A, Seidel T (2002) Renewable energy: current and potential issues. BioScience 52:1111–1120

Pitman RM (2006) Wood ash use in forestry – a review of the environmental impacts. Forestry 79:563–588

Poitrat-Ademe E (1999) The potential of liquid biofuels in France. Renew Energ 16:1084–1089

Postel SL, Daily GC, Ehrlich PR (1996) Human appropriation of renewable fresh water. Science 271:785–788

Pretty JN, Ball AS, Xiaoyun L, Ravindranath NH (2002) The role of sustainable agriculture and renewable resource management in reducing greenhouse-gas emissions and increasing sinks in China and India. Philos Transact A Math Phys Eng Sci 360:1741–1761

Quénéa K, Derenne S, Largeau C, Rumpel C, Mariotti A (2006) Influence of change in land use on the refractory organic macromolecular fraction of a sandy spodosol (Landes de Gascogne, France). Geoderma 136:136–151

Rabbinge R, van Diepen CA (2000) Changes in agriculture and land use in Europe. Eur J Agron 13:85–99

Raich JW, Russell AE, Bedoya-Arrieta R (2007) Lignin and enhanced litter turnover in tree plantations in lowland Costa Rica. Forest Ecol Manag 239:128–135

Ranger J, Turpault M (1999) Input-output nutrient budgets as a diagnostic tool for sustainable forest management. Forest Ecol Manag 122:139–154

Raskin P, Gleick P, Kirshen P, Pontius G, Strzepek K (1997) Water futures: assessment of long-range patterns and problems. Stockholm Environment Institute, Stockholm

Rasse DP, Rumpel C, Dignac M (2005) Is soil carbon mostly root carbon? Mechanisms for a specific stabilisation. Plant Soil 269:341–356

Razafimbelo TL, Barthès B, Larré-Larrouy M, De Luca EF, Laurent J, Cerri CC, Feller C (2006) Effect of sugarcane residue management (mulching versus burning) on organic matter in a clayey oxisol from southern Brazil. Agr Ecosyst Environ 115:285–289

Reicosky DC, Forcella F (1998) Cover crop and soil quality interactions in agroecosystems. J Soil Water Conserv 53:224–229

Rees RM, Bingham IJ, Baddeley JA, Watson CA (2005) The role of plants and land management in sequestering soil carbon in temperate arable and grassland ecosystems. Geoderma 128: 130–154

Reijnders L (2000) A normative strategy for sustainable resource choice and recycling. Resour Conserv Recy 28:121–133

Reijnders L (2003) Loss of living nature and the possibilities for limitation thereof. In: van der Zwaan B, Pedersen A (eds) Sharing the planet. Eburon, Delft

Reijnders L (2005) Disposal, uses and treatments of combustion ashes: a review. Resour Conserv Recy 43:313–336

Reijnders L (2006) Conditions for the sustainability of biomass based fuel use. Energ Policy 34:863–876

Reijnders L (2008) Ethanol production from crop residues and soil organic carbon. Resour Conserv Recy 52:653–658

Reijnders L, Huijbregts MAJ (2003) Choices in calculating life cycle emissions of carbon containing gases associated with forest derived biofuels. J Clean Prod 11:527–532

Reijnders L, Huijbregts MAJ (2005) Life cycle emissions of greenhouse gases associated with burning animal wastes in countries of the European Union. J Clean Prod 13:51–56

Reijnders L, Huijbregts MAJ (2007) Life cycle greenhouse gas emissions, fossil fuel demand and solar energy conversion efficiency in European bioethanol production for automotive purposes. J Clean Prod 15:1806–1812

Reijnders L, Huijbregts MAJ (2008) Palm oil and the emission of carbon-based greenhouse gases. J Clean Prod 16:477–482

Reijneveld A, van Wensem J, Oenema O (2009) Trends in soil organic carbon content of agricultural land in the Netherlands between 1984 and 2004. Geoderma in press

Reimann C, Ottesen RT, Andersson M, Arnoldussen A, Koller F, Englmaier P (2008) Element levels in birch and spruce wood ashes – green energy? Sci Total Environ 393:191–197

Renouf MA, Wegener MK, Nielsen LK (2009) An environmental life cycle assessment comparing Australian sugarcane with US corn and UK sugar beet as producers of sugars for fermentation. Biomass Bioenerg in press

Richardson B, Skinner MF, West G (1999) The role of forest productivity in defining the sustainability of plantation forests in New Zealand. Forest Ecol Manag 122:125–137

Ring E, Jacobson S, Nohrstedt H (2006) Soil-solution chemistry in a coniferous stand after adding wood ash and nitrogen. Can J Forest Res 36:153–163

Robèrt M, Hultén P, Frostell B (2007) Biofuels in the energy transition beyond peak oil. A macroscopic study of energy demand in the Stockholm transport system 2030. Energy 32:2089–2098

Rockström J (2003) Water for food and nature in drought-prone tropics: vapour shift in rain-fed agriculture. Philos T R Soc B 358:1997–2009

Rockström J, Gordon L, Folke C, Falkenmark M, Engwall M (1999) Linkages among water vapor flows, food production, and terrestrial ecosystem services. Conserv Ecol 3(2):[np]

Rockström J, Lannerstad M, Falkenmark M (2007) Assessing the water challenge of a new green revolution in developing countries. P Natl Acad Sci USA 104:6253–6260

Royal Society (2008) Sustainable biofuels: prospects and challenges. http://royalsociety.org

Ryan PJ, Harper RJ, Laffan M, Booth TH, McKenzie NJ (2002) Site assessment for farm forestry in Australia and its relationship to scale, productivity and sustainability. Forest Ecol Manag 171:133–152

Rytter L (2002) Nutrient content in stems of hybrid aspen as affected by tree age and tree size, and nutrient removal with harvest. Biomass Bioenerg 23:13–25

Rytter RM (2005) Water use efficiency, carbon isotope discrimination and biomass production of two sugar beet varieties under well-watered and dry conditions. J Agron Crop Sci 191:426–438

Saffih-Hdadi K, Mary B (2008) Modeling consequences of straw residues export on soil organic carbon. Soil Biol Biochem 40:594–607

Saikku L, Antikainen R, Kauppi PE (2007) Nitrogen and phosphorus in the Finnish energy system, 1900–2003. J Ind Ecol 11(1):103–119

Sanchez PA (1999) Delivering on the promise of agroforestry. Environ Dev Sust 1:275–284

Sanchez PA, Buresh RJ, Leakey RRB (1997) Trees, soils, and food security. Philos T R Soc B 352:949–961

Sander M, Andrén O (1997) Ash from cereal and rape straw used for heat production: liming effect and contents of plant nutrients and heavy metals. Water Air Soil Poll 93:93–108

Sartori F, Lal R, Ebinger MH, Parrish DJ (2006) Potential soil carbon sequestration and $CO_2$ offset by dedicated energy crops in the USA. Crit Rev Plant Sci 25:441–472

Sauerbeck DR (2001) $CO_2$ emissions and C sequestration by agriculture – perspectives and limitations. Nutr Cycl Agroecosys 60:253–266

Searchinger T, Heimlich R, Houghton RA, Dong F, Elobeid A, Fabiosa J, Tokgoz S, Hayes D, Yu T (2008) Use of U.S. croplands for biofuels increases greenhouse gases through emissions from land-use change. Science 319:1238–1240

Sekhon NK, Hira GS, Sidhu AS, Thind SS (2005) Response of soyabean (*Glycine max* Mer.) to wheat straw mulching in different cropping seasons. Soil Use Manage 21:422–426

Shrestha RK, Ladha JK, Gami SK (2006) Total and organic soil carbon in cropping systems of Nepal. Nutr Cycl Agroecosys 75:257–269

Sims REH, Riddell-Black D (1998) Sustainable production of short rotation forest biomass crops using aqueous waste management systems. Biomass Bioenerg 15:75–81

Smeets E, Junginger M, Faaij A, Walter A, Dolzan P, Turkenburg W (2008) The sustainability of Brazilian ethanol – an assessment of the possibilities of certified production. Biomass Bioenerg 32:781–813

Smil V (1991) Population growth and nitrogen: an exploration of a critical existential link. Popul Dev Rev 17:569–601

Smith P (2004) Carbon sequestration in croplands: the potential in Europe and the global context. Eur J Agron 20:229–236

Sofo A, Manfreda S, Fiorentino M, Dichio B, Xiloyannis C (2008) The olive tree: a paradigm for drought tolerance in the Mediterranean climates. Hydrol Earth Syst Sci 12:293–301

Spaccini R, Piccolo A, Haberhauer G, Gerzabek MH (2000) Transformation of organic matter from maize residues into labile and humic fractions of three European soils as revealed by $^{13}C$ distribution and CPMAS-NMR spectra. Eur J Soil Sci 51:583–594

Stape JL, Binkley D, Ryan MG (2008) Production and carbon allocation in a clonal *Eucalyptus* plantation with water and nutrient manipulations. Forest Ecol Manage 255:920–930

Stevens SF (1993) Tourism, change, and continuity in the Mount Everest Region, Nepal. Geogr Rev 83:410–427

Sticklen M (2006) Plant genetic engineering to improve biomass characteristics for biofuels. Curr Opin Biotechnol 17:315–319

Su Z, Zhang J, Wu W, Cai D, Lv J, Jiang G, Huang J, Gao J, Hartmann R, Gabriels D (2007) Effects of conservation tillage practices in winter wheat water-use efficiency and crop yield on the Loess Plateau, China. Agr Water Manage 87:307–314

Sudmeyer RA, Simons JA (2008) *Eucalyptus globulus* agroforestry on deep sands on the southeast coast of Western Australia: the promise and the reality. Agr Ecosyst Environ 127:73–84

Swanston JS, Newton AC (2005) Mixtures of UK wheat as an efficient and environmentally friendly source for bioethanol. J Ind Ecol 9(3):109–126

Swedish Environmental Advisory Council (2007) Scenarios on economic growth and resource demand. Statens Offentliga Utreningar, Stockholm

Syers JK, Powlson DS, Rappaport I, Sanchez PA, Lal R, Greenland DJ, Ingram J (1997) Managing soils for long-term productivity. Philos T R Soc B 352:1011–1021

Tang H, Qiu J, Van Ranst E, Li C (2006) Estimations of soil organic carbon storage in cropland of China based on DNDC model. Geoderma 134:200–206

Tarkalson DD, Payero JO, Hergert GW, Cassman KG (2006) Acidification of soil in a dry land winter wheat–sorghum/corn–fallow rotation in the semiarid U.S. Great Plains. Plant Soil 283: 367–379
Tilman D, Fargione J, Wolff B, D'Antonio C, Dobson A, Howarth R, Schindler D, Schlesinger WH, Simberloff D, Swackhamer D (2001) Forecasting agriculturally driven global environmental change. Science 292:281–284
Tilman D, Cassman KG, Matson PA, Naylor R, Polasky S (2002) Agricultural sustainability and intensive production practices. Nature 418:671–677
Tinker PB (1997) The environmental implications of intensified land use in developing countries. Philos T R Soc B 352:1023–1033
Triberti L, Nastri A, Giordani G, Comellini F, Baldoni G, Toderi G (2008) Can mineral and organic fertilization help sequestrate carbon dioxide in cropland? Eur J Agron 29:13–20
Tricker PJ, Pecchiari M, Bunn SM, Vaccari FP, Peressotti A, Miglietta F, Taylor G (2009) Water use of a bioenergy plantation increases in a future high $CO_2$ world. Biomass Bioenerg in press
Troeh FR, Hobbs JA, Donahue RL (1999) Soil and water conservation: productivity and environmental protection. Prentice Hall, Englewood Cliffs
Tuskan GA (1998) Short-rotation woody crop supply systems in the United States: what do we know and what do we need to know? Biomass Bioenerg 14:307–315
UNEP (2002) Global environment outlook 3: past, present and future perspectives. Earthscan, London
UNESCO (2003) Water for people, water for life. UNESCO and Berghahn Books, Barcelona
Valentin C, Agus F, Alamban R, Boosaner A, Bricquet JP, Chaplot V et al. (2008) Runoff and sediment losses from 27 upland catchments in Southeast Asia: impact of rapid land use changes and conservation practices. Agr Ecosyst Environ 128:225–238
Vance ED (2000) Agricultural site productivity: principles derived from long-term experiments and their implications for intensively managed forests. Forest Ecol Manage 138:369–396
van Dijk AIJM, Keenan RJ (2007) Planted forests and water in perspective. Forest Ecol Manage 251:1–9
Varvel GE, Vogel KP, Mitchell RB, Follett RF, Kimble JM (2008) Comparison of corn and switchgrass on marginal soils for bioenergy. Biomass Bioenerg 32:18–21
Vleeshouwers LM, Verhagen A (2002) Carbon emission and sequestration by agricultural land use: a model study for Europe. Glob Change Biol 8:519–530
Volk TA, Tharakan PJ, Abrahamson LP, White EH (2003) Greater potential for renewable biomass energy. BioScience 53:620–621
Wallace JS (2002) Increasing agricultural water use efficiency to meet future food production. Agr Ecosyst Environ 82:105–119
Webb J, Bellamy P, Loveland PJ, Goodlass G (2003) Crop residue returns and equilibrium soil organic carbon in England and Wales. Soil Sci Soc Am J 67:928–936
Wilhelm WW, Johnson JMF, Hatfield JL, Voorhees WB, Linden DR (2004) Crop and soil productivity response to corn residue removal: a literature review. Agron J 96:1–17
Williams MR, Melack JM (1997) Solute export from forested and partially deforested catchments in the central Amazon. Biogeochemistry 38:67–102
Wood D (1998) Ecological principles in agricultural policy: but which principles? Food Policy 23:371-381
Woodbury PB, Rubin G, McCune DC, Weinstein LH, Neuhauser EF (1999) Assessing trace element uptake by vegetation on a coal fly ash landfill. Water Air Soil Poll 111:271–286
World Water Council (2000) World water vision. Earthscan, London
Worrell R, Hampson A (1997) The influence of some forest operations on the sustainable management of forest soils – a review. Forestry 70:61–85
Wright L (2006) Worldwide commercial development of bioenergy with a focus on energy crop-based projects. Biomass Bioenerg 30:706–714
Wright LL, Hughes EE (1993) US carbon offset potential using biomass energy systems. Water Air Soil Poll 70:483–497

Wright DG, Mullen RW, Thomason WE, Raun WR (2001) Estimated land area increase of agricultural ecosystems to sequester excess atmospheric carbon dioxide. Commun Soil Sci Plant 32:1803–1812

Wynn JG, Bird MI (2007) C4-derived soil organic carbon decomposes faster than its C3 counterpart in mixed C3/C4 soils. Glob Change Biol 13:2206–2217

Zimmerman JB, Mihelcic JR, Smith J (2008) Global stressors on water quality and quantity. Environ Sci Technol 42:4247–4254

Zinn YL, Resck DVS, da Silva JE (2002) Soil organic carbon as affected by afforestation with *Eucalyptus* and *Pinus* in the Cerrado region of Brazil. Forest Ecol Manage 166:285–294

Zinn YL, Lal R, Resck DVS (2005) Changes in soil organic carbon stocks under agriculture in Brazil. Soil Till Res 84:28–40

Zwart SJ, Bastiaanssen WGM (2004) Review of measured crop water productivity values for irrigated wheat, rice, cotton and maize. Agr Water Manag 69:115–133

Zwart K, Oudendag P, Kuikman P (2007) Co-digestion of animal manure and maize: is it sustainable? Alterra, Wageningen

# Chapter 4
# Climate Effects and Non-greenhouse Gas Emissions Associated with Transport Biofuel Life Cycles

## 4.1 Introduction

Here we will consider climate effects and non-greenhouse gas emissions associated with the life cycles of transport biofuels. Considering the whole seed-to-wheel life cycle of transport biofuels is important because at each stage, there may be significant environmental impacts. In cropping plants, there are usually inputs of fossil fuels (needed to power tractors and generate N fertilizer) and emissions of substances such as $N_2O$ and nitrate (both derived from nitrogen fertilizers) and pesticides. The effects thereof may be substantial. For instance, Donner and Kucharik (2008) have shown that a US bioethanol production of 15 billion gallons would increase the annual average flux of dissolved inorganic fixed nitrogen (N) in the Mississippi and Atchafalaya rivers by 10–34%. And it has been argued that in Brazil, ethanol production from sugar cane will contribute to a rapidly changing tropical biogeochemical N cycle, because the N fertilizer use efficiency of production is low: about 30% of N fertilizer ends up in sugar cane tissue (Galloway et al. 2008).

Also the choice of plants grown as biofuel feedstocks may matter, as plants, for instance, differ in their production of isoprene, which may contribute oxidizing smog (Royal Society 2008). The stage of processing biomass to transport biofuels is associated with energy use and process-related emissions. For instance, dry mill fuel ethanol production leads to significant emissions of ethanol, acetaldehyde, acetic acid and ethylacetate (Brady and Pratt 2007). Leakage from storage facilities for bioethyl-*tertiary*-butylether (ETBE) may have a significant impact on groundwater quality (Rosell et al. 2007). And the stage of driving is important, because driving a car on biofuels generates emissions which may be different from those of fossil fuels.

This chapter starts with an overview of the different types of uncertainty related to life cycle studies (Sect. 4.2). After that, most attention will be given to life cycle emissions of biofuels based on terrestrial plants and wastes. These are considered in Sects. 4.3 to 4.5. Section 4.6 will deal with possibilities to reduce the life cycle emissions associated with the biofuels considered here.

L. Reijnders, M.A.J. Huibregts, *Biofuls for Road Transport*

## 4.2 Uncertainty in the Life Cycle Environmental Impact Assessments of Biofuels

Estimates of life cycle impacts are subject to uncertainty. In life cycle assessments, there is uncertainty in input data (parameter uncertainty), in normative choices (scenario uncertainty) and mathematical relationships (model uncertainty) (Huijbregts et al. 2003; Lloyd and Ries 2007). As the focus in this chapter is largely on comparing fuels, model uncertainty tends to be rather similar for all fuels, which is favourable to the comparative value of life cycle assessments.

Parameter uncertainty may be limited by using relatively good quality inventories of emission and resource use data, such as the JLCA-LCA inventory from Japan (Suguiyama et al. 2005) and the Ecoinvent database (cf. Zah et al. 2007), as well as recent peer-reviewed research into emissions and resource use. In this way, uncertainties in contributions of industrial and transport activities to impacts of biofuels, especially in industrialized countries, can be limited. However, uncertainties about industrial activities in some developing countries may remain relatively large because of major uncertainty about fuels, energy efficiency and environmental technology (e.g. Reijnders and Huijbregts 2008a; de Vries 2008). Uncertainties linked to the fate of C and N in cropping and harvesting feedstocks are relatively large. In the case of $N_2O$ emissions associated with intensive cropping, uncertainty in net greenhouse gas emissions may well be $\pm 20\%$ (Reijnders and Huijbregts 2008b), and uncertainties about changes in C stocks of ecosystems may also be quite substantial.

Normative choices are an important source of uncertainty. One of these choices relates to the time that land will be used to produce biofuel. Choices regarding this time are important in calculating net greenhouse gas emissions due to land use change (e.g. Reijnders and Huijbregts 2008a; Wicke et al. 2009). Allocation in the case of multi-output production is another normative choice that is important. As explained in Chap. 2, there are three major ways to allocate. The first is based on prices, the second on physical categories, such as energy or weight, and the third on subtracting avoided processes (also called substitution). Apart from choosing the basis for allocation, there may also be other matters to consider. Take, for example, the conversion of lignocellulose in dried distillers grains with or without solubles [DDG(S)] to ethanol. Lignocellulosic outputs of ethanol production such as dried distillers grains are currently an ingredient of animal feed (Taylor-Pickard 2008). If one goes back in history, there have been advocates for replacing ingredients that had high starch contents by lignocellulosic ingredients, such as DDG(S), in animal feed (e.g. Sarkanen 1976). Now, when these lignocellulosic components are diverted to the production of transport fuel, will this give rise to an increase in the starch content of animal feed? And should the effects thereof on the net emission of greenhouse gases be allocated to lignocellulosic ethanol, and if so to what extent? Also, an objection has been raised against considering dried distillers grains as a product output of ethanol production, to which non-product outputs should be allocated (Patzek 2004). According to Patzek (2006) dried distillers grains should become an input in cropping and spread on the fields to diminish the need for nitrogen fertilizer, decrease soil erosion and improve the energy efficiency of cropping. Patzek (2006) has also argued that if there is any crediting at all of dried distillers grains, the energy credit should be somewhat negative.

The complications of dealing with co-products such as DDG(S) and allocation may well seem so problematic that no 'iron-clad' estimate of net greenhouse gas emissions associated with biofuels from multi-output processes seems feasible. Only limited study has been made as to the differences in estimated environmental impacts of transport biofuels following from the different approaches to allocation. Eickhout et al. (2008) found that the substitution approach and allocation on the basis of energy-generated outcomes for ethanol and biodiesel were in the range of $\pm 15\%$. Curran (2007) looked at the impact of different ways of allocation (based on price, weight, volume and energy) on the relative environmental ranking of conventional gasoline and bioethanol and found that this ranking was the same in all instances. On the other hand, Reijnders and Huijbregts (2005) found that allocation based on either price or energy may lead to a difference in the environmental ranking of fossil-fuel- and manure-based electricity. And Malça and Freire (2006) did show that in the case of wheat ethanol, different ways to allocate have a major influence on results.

In the following, we will indicate what type of time frame and allocation has been used in arriving at specific results.

# 4.3 Transport Biofuels and Climate

## *4.3.1 Introduction*

Transport biofuels made from plants are often called 'climate neutral' or 'carbon neutral'. These terms can be traced back to the participation of plants in the biogeochemical C cycle. Plants take up $CO_2$ from the atmosphere and convert this into biomass, and when biomass is burned, the $CO_2$ is 'given back' to the atmosphere. There is said to be C neutrality: over a short time span, sequestration equals emission of $CO_2$, which is a greenhouse gas. Greenhouse gases are transparent for relatively high energy solar radiation, such as visible light, but absorb infrared radiation and thereby influence atmospheric temperature. And thus, carbon neutrality is in this case said to equal climate neutrality. However, there is more to the relation between biofuels and climate. In part, this is linked to the direct effect of plants on local climate and in part to the emission of non-$CO_2$ greenhouse gases, such as $N_2O$ and $CH_4$. Also, biofuel production can be accompanied by changes in C sequestration by ecosystems.

If compared with the original 'natural' vegetation, cropping biofuels may differ in determinants of local climate, such as surface roughness (Notaro et al. 2006), evapo(tanspi)ration (Gustafsson et al. 2004; McPherson 2007), precipitation (Liu et al. 2006) and albedo (Gustafsson et al. 2004; Schneider et al. 2004; McPherson 2007). Albedo is a measure of the reflection of solar radiation by the earth's surface (including vegetation), which in turn is a determinant of net radiation. Net effects of vegetation change may be different dependent on region. In cold regions, replacement of forest by annual biofuel crops tends to have a cooling effect, due to the importance of change in albedo, and in tropical regions, this replacement may cause warming, mainly due to a decrease in evaporation and cloud cover (Betts et al. 2007). When changes in vegetation are widespread, there may be knock-on effects on climate on a wider scale (Delire et al. 2001; Liu et al. 2006; Betts 2007; Betts et al. 2007; McPherson 2007). These will be further discussed in Chap. 5.

Here, of the factors that may impact climate, we will only further consider net greenhouse gas emissions. These may be positive or negative. The latter case corresponds with net C sequestration. First, we will consider the major determinants of these net emissions. Thereafter, the actual net greenhouse gas emissions of a number of transport biofuels will be considered.

Potentially important determinants of the net greenhouse gas emissions linked to the transport biofuel life cycle are:

- Carbonaceous greenhouse gas emissions linked to the cumulative demand for fossil fuels.
- $N_2O$ emissions linked to N inputs in, and non-product outputs (e.g. $NO_x$ emissions) of, biofuel production.
- Changes in atmospheric $CO_2$ concentrations following from changes in carbon sequestration. The latter may relate to changes in soil carbon level and/or changes in aboveground biomass.
- Emissions of biogenic, non-$CO_2$ carbonaceous greenhouse gases linked to the biofuel life cycle. These include $CH_4$ emissions linked to anaerobic conversion of biomass and non-$CO_2$ carbonaceous greenhouse gas emissions due to biomass burning.

These determinants will be considered in turn.

### *4.3.2 Fossil-Fuel-Based Carbonaceous Greenhouse Gas Emissions*

Transport biofuels replace fossil fuels. But because, as pointed out in Chap. 2, much production in current societies is dependent on fossil fuels, it will come as no surprise that burning fossil fuels is often an important contributor to the greenhouse gas emissions associated with biofuels. N fertilizers are often made on the basis of natural gas; tractors and transport are often powered by fossil fuels based on mineral oil. Factories, involved in converting biomass into fuels fit for powering vehicles, are more mixed in their fuel use. There are production facilities doing without burning fossil fuels. In Brazil, factories converting sugar into ethanol are often powered by harvest residues of sugar cane (Macedo et al. 2008). In Sweden, biofuel production in factories tends to use wood chips from forest logging residues (Börjesson and Mattiasson 2008). But, for example, in France and Germany, factories producing bioethanol or biodiesel are usually powered with fossil fuels (Reijnders and Huijbregts 2007, 2008b). Life cycle assessments of biofuels are characterized by a long-standing interest in the $CO_2$ emissions linked to cumulative fossil energy demand (e.g. Pimentel et al. 1973; Weisz and Marshall 1979), and by now the standard is that they include the $CO_2$ emission linked to burning fossil fuels to power the biofuel life cycle, though there are still exceptions to this rule (e.g. Chisti 2007, 2008; Dismukes et al. 2008).

There may also be non-$CO_2$ carbonaceous emissions linked to fossil fuel use. For instance, there may be leakage of $CH_4$ (methane) during transport and use of natural gas, and, if burning is not optimized, there may be emission of hydrocarbons and carbon monoxide (CO). The non-$CO_2$ carbonaceous gases on a molecule-for-molecule basis tend to have a greater greenhouse effect than $CO_2$. The non-$CO_2$ carbonaceous greenhouse gases are often neglected in life cycle assessments, which will lead to an underestimate of the greenhouse effect of actual emissions. However, in advanced industrial economies, the error linked to this underestimation will be small, in the order of a few percent maximally.

### *4.3.3 $N_2O$ Emissions*

$N_2O$ emissions may originate on several occasions along the biofuel life cycle (Reijnders and Huijbregts 2005). The production of N fertilizers is often accompanied by the emission of $N_2O$. $NO_x$ emissions linked to burning fossil fuels may be deposited as N compounds in soils, and there, they may be partly converted by microorganisms into $N_2O$. N inputs in cropping are also partly converted into $N_2O$. It is usually assumed that the latter process is, directly and indirectly, responsible for most of the $N_2O$ emission linked to the transport biofuel life cycle. The actual quantity of the $N_2O$ emission, given a specified input of N and soil, is, however, the subject of a lively debate (Mosier et al. 1998; Crutzen et al. 2007). According to the estimates of Crutzen et al. (2007), 3–5% of the N input into growing biofuel crops will be converted into $N_2O$. Mosier et al. (1998) have presented data suggesting that direct $N_2O$ emissions from agricultural fields associated with biofuel cropping may be about 1.25% of added fixed nitrogen. In addition, they argue that fixed nitrogen lost from agricultural fields may also be subject to microbial conversion to $N_2O$ (estimated at 2.5% of fixed N lost).

Local conditions may have a significant impact on actual $N_2O$ emissions. The presence of soil moisture matters (Rebelo de Mira and Kroeze 2006; Guo and Zhou 2007; Scheer et al. 2008). Higher temperatures tend to be conducive to higher emissions of $N_2O$ (Ding et al. 2007; Scheer et al. 2008). So are higher levels of soil organic carbon (Guo and Zhou 2007; Liu et al. 2007). In some soils, nitrate is relatively favourable to $N_2O$ production, but in other soils, it is rather ammonia (Guo and Zhou 2007; Liu et al. 2007; Scheer et al. 2008). Moreover, it may be noted that $N_2O$ emissions may change when the climate changes. Temperature and precipitation are significant determinants of $N_2O$ emissions, and as temperature and precipitation are expected to change when the climate changes, $N_2O$ emissions from biofuel cropping may be different in the future from what they are now (Novoa and Tejeda 2006). In view of variability, there is a case for a rather wide range for the conversion of fixed N into $N_2O$: 1.5–5%.

### *4.3.4 Biogenic $CO_2$ Emissions*

The production of transport biofuels can be accompanied by changes in carbon sequestration. Firstly, there can be changes in the carbon content of ecosystems. There may be both losses from, and increases of, C in the ecosystem, which in turn will change atmospheric $CO_2$ concentrations. In early life cycle assessments, quantification of such changes was largely neglected (with some exceptions, such as Reijnders and Huijbregts 2003, Delucchi 2005; Kim and Dale 2005 and Cowie et al. 2006). However, by now, changes in C sequestration by ecosystems are increasingly recognized as a major determinant of net seed-to-wheel greenhouse gas emissions (e.g. Fritsche 2007; Danielsen et al. 2008; Fargione et al. 2008; Gibbs et al. 2008; Nechodom et al. 2008; Searchinger et al. 2008; Wicke et al. 2009). In a number of cases, the link between change in C sequestration and expansion of biofuel production is *direct*. For instance, because forest is cleared or degraded land is planted for oil palm plantations serving the biodiesel market. In these cases, there is a large effect on C sequestration (Danielsen et al. 2008; Fargione et al. 2008). There are also cases in which land use change has more limited direct effects on C stocks – for instance, cases where oil palms have replaced other plantation crops, such as rubber or coconut, that gave lower revenues (Tan et al. 2009).

There may also be *indirect* effects of expanding biofuel production on land use. These follow from the relative inelasticity of demand for food. When land used for food or feed (fodder) production is diverted to use for biofuel production, one may expect that food or feed production to a very large extent moves elsewhere (Searchinger et al. 2008). Some examples may illustrate this. Currently, in Brazil, there is substantial conversion of pasture to arable land to grow soybeans. Conversion of pasture to arable land tends to lower carbon in the ecosystem. But this is not the 'whole story'. Farmers that used the pasture may move to new pastures or may import fodder from elsewhere, for which deforestation may be necessary (Nepstad et al. 2008). Also, the expansion of soybean cropping in Brazil is partly linked to the expansion of corn-for-biofuel production in the United States, which has displaced US soybean production (Nepstad et al. 2008; Scharlemann and Laurance 2008). Tilman et al. (2006) have proposed the use of native prairie species as a basis for transport biofuel production from the US prairies, but to the extent that such production replaces current grazing by livestock, the latter has to be accommodated somewhere else. A last example refers to European wheat. In 2007, the price thereof was much elevated, in part due to the use of wheat starch for ethanol biofuel production. This led to a massive rise of corn imports with Brazil as a major supplier, which in turn was conducive to change of Brazilian land with natural vegetation into cropland.

There may also be other changes in carbon sequestration linked to biofuel life cycles. For instance, both in the case of electricity production of biomass and the production of synthesis gas from biomass, $CO_2$ may be captured and stored in soils (e.g. aquifers or abandoned gas fields). In the case of electricity production, depending on the technology used, life cycle emissions of greenhouse gases linked to electricity production may be reduced by 75–84%, though other emissions such as those of eutrophying and acidifying substances may be increased (Odeh and Cockerill 2008). There is currently only limited application of such capture and storage (abbreviated CCS), but this may change in the future (Odeh and Cockerill 2008; Gibbins and Chalmers 2008). Also, in generating transport biofuels from biomass by pyrolysis, significant amounts of charcoal, also called 'biochar', may be generated (Demirbaş 2001), which in turn may be added to soils. This approach is currently not practiced, but again, this may change in the future (Marris 2006; Renner 2007).

Changes in carbon sequestration in ecosystems are currently common in transport biofuel production. Clearing forests to allow directly or indirectly for cultivation of transport biofuel crops corresponds with substantial emissions of greenhouse gases (Fearnside and Laurance 2004; Righelato and Spracklen 2007; Fargione et al. 2008; Reijnders and Huijbregts 2008a; Searchinger et al. 2008). This follows to a large extent from the large changes in aboveground and soil carbon stocks. For instance, estimates of aboveground C stocks in rainforests range from approximately 130–270 Mg C $ha^{-1}$ (Danielsen et al. 2008; Fargione et al. 2008; Reijnders and Huijbregts 2008a), and similar estimates for aboveground C stocks in temperate forests range from approximately 100–160 Mg $ha^{-1}$ (Searchinger et al. 2008). Clearing other types of natural vegetation for transport biofuel production may lead to a lower but still substantial desequestration of C. For instance, clearing Cerrado savannah for soybean biodiesel production leads to an average desequestration of about 23 Mg C $ha^{-1}$ (Reijnders and Huijbregts 2008b), whereas losses of C stocks in soil due to conversion of US grasslands into cropland for biofuels have been estimated at approximately 17–40 Mg C $ha^{-1}$ (Fargione et al. 2008). On the other hand, Germer and Sauerborn (2007) have suggested that planting oil palms on degraded grassland may lead to a substantial accumulation of aboveground and soil organic carbon, estimated at 135 Mg C $ha^{-1}$ over a period of 25 years. Nechodom et al. (2008) have suggested very favourable life cycle assessments regarding biomass from forest remediation in California, as it is assumed that without using such biomass, forest fire risk would be much higher.

If one takes account of changes in carbon sequestration due to ecosystem changes in life cycle environmental impact assessments, the distribution of changes in the carbon content linked to land use change over the subsequent period should be established, as pointed out in Sect. 4.2. The Intergovernmental Panel on Climate Change (IPCC 2006) has suggested the use of a 20-year period for this purpose, but calculations also have been made for periods of up to 100 years (Reijnders and Huijbregts 2008a; Wicke et al. 2009).

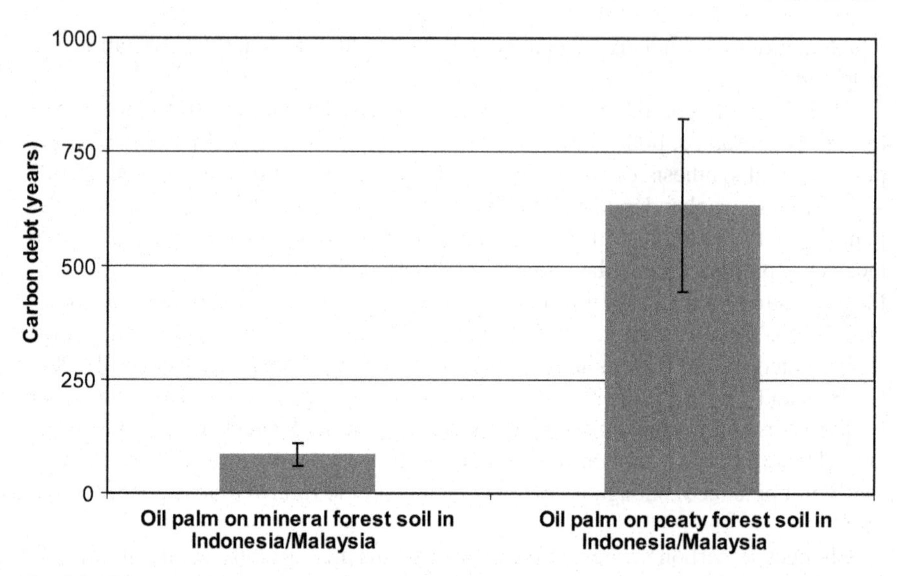


**Fig. 4.1** Time needed (in years) to pay off the carbon debt due to ecosystem C losses in palm oil production according to Fargione et al. (2008) and Danielsen et al. (2008)

Another way to approach this matter is to balance the net reduction of C emissions from the use of biofuels against the 'biofuel carbon debt' due to C losses from ecosystems. This gives rise to a number of years to pay off the carbon debt. An example thereof for palm oil, reflecting a large effect on C stocks, is given in Fig. 4.1.

Direct changes in land use linked to expanded transport biofuel production are relatively easy to include in life cycle assessments. Including indirect changes is more difficult. A first possibility to take account of changes in C sequestration due to indirect changes in land use has been suggested by Fritsche (2007). Fritsche proposed that for such indirect effects, there should be a 'risk adder' or 'iLUC', reflecting the risk that there may be clearance of natural vegetation or other forms of land use change that affect C sequestration and the net emission of $CO_2$ (usually distributed over a 20-year period in line with recommendations of the Intergovernmental Panel on Climate Change). The impact of emissions associated with different assumptions as to indirect effects on land use change (different risk adders or iLUCs) may be considerable. Illustrations are given by Table 4.1.

A more sophisticated approach uses modelling to estimate land use and land use change induced by expanding transport biofuel production. A proposal for such modelling has been published by Kløverpris et al. (2008). Searchinger et al. (2008) have applied this type of modelling using a representation of outcomes in terms of carbon debt. Searchinger et al. (2008) evaluated expanding corn cropping to achieve the US government goals set for the transport biofuel supply in 2016. They used a model to estimate worldwide land use changes following from this development, which, they estimate, will divert 12.8 million ha of US cropland to transport biofuel production. Searchinger et al. (2008) estimate that this will bring 10.8 million

**Table 4.1** Impact of variable 'risk adders' or 'iLUCs' (assumptions as to land use change) on net greenhouse gas emissions if compared with fossil fuels (based on data from Bergsma 2007 and Fritsche 2007; allocation on the basis of prices)

| Biofuel | Plant | Fossil reference (100%) | $CO_2$ equivalent emission if compared with fossil fuel, while including 25% induced deforestation (%) | $CO_2$ equivalent emission if compared with fossil fuel, while including 75% induced deforestation (%) |
|---|---|---|---|---|
| Biodiesel | Rapeseed (Europe) | Diesel | 113 | 241 |
| Biodiesel | Palm oil (Indonesia) | Diesel | 121 | 210 |
| Bioethanol | Wheat (Europe) | Petrol | 100 | 208 |
| Fischer–Tropsch diesel | Lignocellulosic biomass (Europe) | Diesel | 36 | 86 |

ha of additional land into cultivation. Searchinger et al. (2008) have calculated that over a 30-year period, including land use change linked to expanding US corn-based ethanol production will add about 93% to greenhouse gas emissions, if compared with fossil gasoline. The payback time for the carbon debt caused by land use change by corn ethanol is estimated by Searchinger et al. (2008) at 163 years. Modelling such as performed by Searchinger et al. (2008) is dependent on assumptions. Important among these are assumptions regarding the future productivity of land. Searchinger et al. (2008) assumed per hectare an average worldwide corn yield increase of 11.5% between 2007 and 2016. A lower increase would lead to a larger indirect effect on land use, a higher increase to a lower indirect effect. For instance, an additional 20% increase in yield per hectare would decrease the carbon debt from 163 to 133 years, whereas a yield increase of 1.5% between 2007 and 2016 would increase the carbon debt to 183 years (Searchinger et al. 2008).

Thus, land use change may have a very large impact on seed-to-wheel greenhouse gas emissions. Biofuel crops may, furthermore, differ in their consequences for carbon stocks in the soil on which they are grown after land use change. Annual cropping of European arable land has been associated with losses of, on average, 0.84 $Mg\,ha^{-1}\,year^{-1}$ soil carbon from the upper soil layer (Vleeshouwers and Verhagen 2002), and this has implications for the net greenhouse gas emissions associated with biofuel crops such as rapeseed and wheat (Reijnders and Huijbregts 2007, 2008b). Net losses of soil carbon from arable soils under annual crops are not restricted to Europe. In Eastern Canada, arable land on average loses 0.07 Mg C $ha^{-1}\,year^{-1}$ to the atmosphere (Gregorich et al. 2005). In China, the estimated loss on average is 0.81 Mg C $ha^{-1}\,year^{-1}$ (Tang et al. 2006). The loss of soil organic carbon in China parallels the removal of approximately 300 million tons of straw (Wright 2006). From Nepal, losses of soil carbon have been noted in a variety of cropping systems, with, for instance, losses in maize/millet cropping systems ranging between 0.11 and 0.23 Mg C $ha^{-1}$ $year^{-1}$ (Matthews and Pilbeam 2005; Shrestha et al. 2006). In a set of 14 cropped Brazilian soils sampled annually, losses of soil organic carbon were on average 0.15 Mg $ha^{-1}$ $year^{-1}$ (Zinn et al. 2005), whereas for soybean-based crop rotation in the Brazilian Cerrado region, losses were reported to be between 0.5 and 1.5 Mg C $year^{-1}$ $ha^{-1}$ (Jantalia et al. 2007). In Africa, losses of carbon from soil are expected as little or no agricultural residues are returned to soils in many cropping systems (Syers et al. 1997). In the case of semi-arid Sudan, an annual loss from cropland during the twentieth century of, on average, 0.29 Mg C $ha^{-1}$ $year^{-1}$ has been found (Ardo and Olsson 2003). Long-term experiments in East Africa have suggested that losses of 0.69 Mg C $ha^{-1}$ $year^{-1}$ are common (Nandwa 2001). Over relatively short and recent time spans, measured C losses in the southern Ethiopian highlands amounted to approximately 0.85–1.75 Mg C $ha^{-1}$ $year^{-1}$ (Lemenih and Itanna 2004), and in Western Burkina Faso to approximately 0.31 Mg C $ha^{-1}$ $year^{-1}$ (Ouattara et al. 2006).

From peaty arable soils, losses may be much higher. They are especially high when there is deep drainage and intensive mechanical soil disturbance (Freibauer et al. 2004). Net carbon losses varying from 6 Mg in northern Norway up to 15 Mg C $ha^{-1}$ $year^{-1}$ in the tropics have been reported (Grønlund et al. 2006; Reijnders and Huijbregts 2007, 2008a).

Mechanical tillage and deep ploughing tend to favour net losses of soil carbon stocks (Fontaine et al. 2007). No-till practices, ceteris paribus, lead to higher levels of soil carbon than tillage (Fontaine et al. 2007), so does the use of cover crops (Pretty et al. 2002). Also, returning harvest residues to soil is conducive to carbon sequestration in soils (Reijnders and Huijbregts 2007; Lal 2008). In temperate climates, soil tillage can be combined with a stable level of soil carbon, when fresh inputs of C (e.g. manure, crop residues) are large enough (Reijneveld et al. 2009).

Whereas losses of soil carbon are not uncommon under annual crops, net carbon sequestration can occur when perennials or multiannual crops are grown.

Growing switchgrass that may serve as a cellulosic feedstock for transport biofuel production is associated with the accumulation of soil carbon in the upper soil layer (Ma et al. 2000; Lal 2008). Carbon accumulation rates depend on climate and soil and time period chosen. McLaughlin and Adams Kszos (2005) found in the mid-Atlantic region of the USA over a 6-year period an accumulation of 1.2–1.6 Mg C $ha^{-1}$ $year^{-1}$. A simulation study considering a 30-year period of switchgrass cropping in the Eastern USA suggests a carbon accumulation rate of 0.53 Mg $ha^{-1}$ $year^{-1}$ (McLaughlin et al. 2002). Long-term trials with ley grass in Sweden suggested an annual accumulation of 1.0–1.3 Mg C $ha^{-1}$ $year^{-1}$ over a 30-year period (Börjesson 1999). Björesson (1999) has suggested that a switch from annual crops to perennial biofuel crops may be associated with a gain of about 0.5 Mg C $ha^{-1}$ $year^{-1}$ in Swedish mineral soils. However, net sequestration does not always occur under perennial crops. In a grassland with *Miscanthus sinensis* in Japan, harvested yearly, reductions of carbon stocks were observed in the order of 0.56–1 Mg $ha^{-1}$ $year^{-1}$ (Yazaki et al. 2004). Finally, soil organic carbon levels may change when the climate changes. The effects of climatic change are complex and may be different dependent on geography and the (agro)ecosystem (Luo 2007)

For Europe, Vleeshouwers and Verhagen (2002) estimated that an increase in average temperature of 1 °C may be associated with an average net loss of soil organic carbon in arable soils of about 0.04 Mg ha$^{-1}$ year$^{-1}$.

### *4.3.5 Other Carbonaceous Biogenic Emissions*

When vegetation is cleared to make way for biofuel cropping, there may be significant emissions of non-$CO_2$ carbonaceous greenhouse gases, including CO and hydrocarbons. In practice, these gases may add, in $CO_2$ equivalents, 10–20% to the emission of $CO_2$ only (Reijnders and Huijbregts 2008a). Also, if compared with forested land, arable land with annual biofuel crops becomes a reduced sink for $CH_4$ (Powlson et al. 1997). Long-term cultivation has been shown to reduce $CH_4$ uptake and oxidation by soils by 85% in a temperate setting. This may correspond to a reduction of the soil sink for $CH_4$ in the order of 100–200 kg $CO_2$ equivalent ha$^{-1}$ year$^{-1}$. And the nature of cultivation also matters, with synthetic $NH_4$ fertilizer completely inhibiting $CH_4$ oxidation, whereas manure has no inhibitory effect (Powlson et al. 1997). When organic wastes of biofuel production are anaerobically converted in open ponds or in dumps, there may be very large emissions of methane (Reijnders and Huijbregts 2008a).

## 4.4 Net Greenhouse Gas Emissions for Specific Biofuels and Categories of Biofuels

The data presented in Sect. 4.3 for the four major determinants of net greenhouse gas emissions will be used in this section to calculate life cycle emissions for specific biofuels and categories of transport biofuels.

### *4.4.1 Net Greenhouse Gas Emissions for Specific Biofuels*

**Example 1: Electricity from Woody Biomass Instead of from Fossil Fuels**

Substituting fossil fuels with woody biomass from sustainably managed forests in electricity production will strongly reduce the life cycle greenhouse gas emissions per kW h$_e$, as indicated by Fig. 4.2.

However, when there are changes in C sequestration due to forestry, emissions may deviate in a major way from the values in the last column of Fig. 4.2. Nechodom et al. (2008) have given a far more favourable assessment of electricity generated by burning woody biomass from forest remediation in California, as such remediation is supposed to reduce C losses due to fires, whereas use of woody biomass associated with clear cutting forests will lead to greenhouse gas emissions that are much

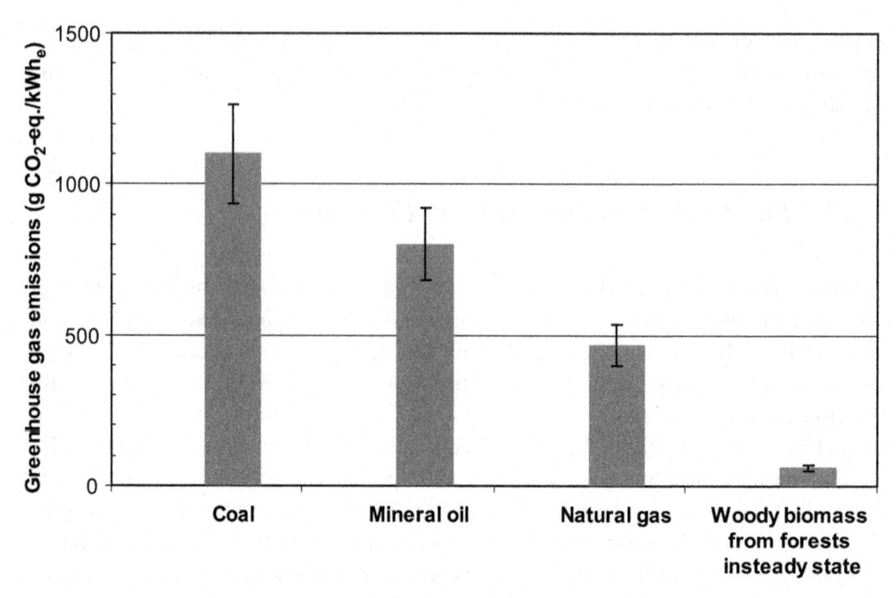


**Fig. 4.2** Life cycle emission of greenhouse gases for different types of electricity production (Reijnders and Huijbregts 2003; Weisser 2007)

larger than those associated with the fossil fuels given in Table 4.1 (Reijnders and Huijbregts 2003). There may also be temporal deviations from the woody biomass value given in Fig. 4.2 when the requirement is that cutting trees should be balanced by planting new trees. In this case, there may be a period of up to 20–40 years after harvesting in which forests are net sources of atmospheric carbon, and even longer periods before initial C losses are fully compensated (Reijnders and Huijbregts 2003).

### Example 2: Ethanol from European Wheat

When it is assumed that straw should be returned to a large extent to soils, the input of fossil fuels in the life cycle of bioethanol from European wheat is such that, whatever the allocation between ethanol and co-products [DDG(S) and glycerol], there is a rather small margin between the fossil-fuel-based energetic input and energetic output of the life cycle (von Blottnitz and Curran 2007; Reijnders and Huijbregts 2007). This limited advantage regarding the emission of greenhouse gases for bioethanol is, however, well exceeded by the average emissions $N_2O$ and $CO_2$ from conventionally tilled soils on which wheat is grown, whatever the allocation chosen (Crutzen et al. 20007; Reijnders and Huijbregts 2007). When indirect effects on land use are also factored in, ethanol from European wheat does worse than fossil gasoline, whatever the choices regarding allocation (Fritsche 2007; Reijnders and Huijbregts 2007).

**Example 3: Palm Oil**

Palm oil has been proposed for use as a heavy-duty transport fuel in warm climates (Prateepchaikul and Apichato 2003). It may also be used in electricity production and after transesterification as biodiesel in cars (Reijnders and Huijbregts 2008a). In the case of palm oil, there is, both in monetary and physical terms, one major marketable output of production (palm oil) and one minor one: palm kernel cake (which may be used as animal feed). For the establishment of oil palm plantations, a variety of land use changes is possible, and the effects thereof can be distributed over time in different ways (Reijnders and Huijbregts 2008a; Wicke et al. 2009). It turns out that when the plantation is on mineral soil and native forest is cleared, palm oil will do worse than diesel based on mineral oil unless the changes in C content of the ecosystem are distributed over a long time span (Reijnders and Huijbregts 2008a; Wicke et al. 2009). When the plantation is on peaty soil, palm oil does much worse regarding the emission of carbonaceous greenhouse gases and $N_2O$ than diesel based on mineral oil (Danielsen et al. 2008; Fargione et al. 2008; Reijnders and Huijbregts 2008a; Wicke et al. 2009). On the other hand, when a palm oil plantation is established on abandoned mineral soils, palm oil will do much better than fossil diesel (Germer and Sauerborn 2007; Wicke et al. 2009).

### *4.4.2 Net Greenhouse Gas Emissions for Categories of Transport Biofuels*

On the basis of available data, it is also possible to draw general conclusions as to the question of whether categories of transport biofuels are better or worse regarding life cycle greenhouse gas emissions than conventional fossil transport fuels, such as diesel and petrol. The latter have a life cycle greenhouse gas emission of about 3.6 kg $CO_2$ equivalent $kg^{-1}$ fuel (EUCAR et al. 2007; Reijnders and Huijbregts 2008b). In the following, comparison between fuels will be made on the basis of equal energy generation.

**Transport Biofuels from Crops on Peat**

Transport biofuels from crops grown on peat land do badly as to greenhouse gas emissions. This is linked to the large carbon losses from peat land when used for cultivation. This has been shown for bioethanol in Europe (Reijnders and Huijbregts 2007) and for palm oil from Southeast Asia (Danielsen et al. 2008; Fargione et al. 2008; Reijnders and Huijbregts 2008a). This will also hold for produce of the sago palm, which often grows on peat land (Melling et al. 2005a, b; Singhal et al. 2008). Sago has been called the 'starch crop of the twenty-first century' and suitable for the production of the transport biofuel ethanol (Singhal et al. 2008). The yield of sago plantations is 2–3 Mg starch $year^{-1}$ $ha^{-1}$ (approximately 0.6–0.9 Mg C) (Singhal et al. 2008), but the annual loss of C from peaty soil is approximately 11 Mg C $year^{-1} ha^{-1}$, partly as methane (Melling et al. 2005a, b). When mineral soils have a high carbon content, as in the case of peaty clay, net greenhouse gas emissions will also be relatively high (Reijneveld et al. 2009), making transport biofuel production on such soils relatively unattractive for mitigating climate change.

### Transport Biofuels Derived from Annual Crops on Mineral Soils

Biodiesel made from vegetable oil which may also serve as food or feed is usually inferior to conventional diesel, when the arable soil is subject to tillage without addition of large amount of fresh C (e.g. residues, manure) and land use changes are factored in. Ethanol from starch and sugar crops is usually inferior to conventional gasoline when arable land is subject to tillage, ethanol production is powered by fossil fuels, land use change is factored in and Intergovernmental Panel on Climate Change (IPCC) guidelines are applied. It has been argued that no foreseeable changes in agricultural or energy technology will be able to achieve meaningful benefits as to the emission of greenhouse gases if annual crop-based biofuels are produced at the expense of tropical forests (Gibbs et al. 2008).

Among the transport biofuels from sugar and starch crops, ethanol from sugar cane does relatively well. For Brazilian sugar-cane-derived ethanol produced in 2005/2006, a greenhouse gas emission has been published of somewhat more than 0.4 kg CO equivalent per litre ethanol (Macedo et al. 2008). The latter energetically equals about 0.75 l gasoline, which has a life cycle $CO_2$ emission of about 2.5 kg $CO_2$ equivalent. The estimate of Macedo et al. (2008) includes an $N_2O$ emission from soils, though linked to a lower input of N fertilizer than used by Machado et al. (2008), and a greenhouse gas emission linked to burning ‘trash’. Changes in aboveground and belowground C were not accounted for by Macedo et al. (2008), and the value for the emission of $N_2O$ used by them was well below the 3–5% of N input suggested by Crutzen et al. (2007).

There is no clarity regarding the impact of sugar cane cultivation on soil C stocks. It has been suggested that a conversion of rainforest to pasture and then to sugar cane plantation reduced the soil C stock by about 40% (Groom et al. 2008). But there are also reports that current practices are not associated with reductions in soil C of current arable soils with limited tillage (La Scala et al. 2006; de Resende et al. 2006). However, a major loss of carbon is associated with converting the wooded Cerrado to arable land for growing sugar cane. Fargione et al. (2008) estimate that it will take 17 years to pay back the carbon debt for this by producing sugar cane ethanol. If the IPCC guidelines are followed, ethanol from sugar cane for which the Cerrado savannah has been cleared would probably do somewhat better than fossil gasoline. If arable land remains in use for many decades, sugar cane ethanol will do substantially better than conventional gasoline.

### Transport Biofuels Derived from Perennials

Lignocellulosic biomass produced in a way that does not significantly change C sequestration in ecosystems will do better than fossil fuels in electricity production (see Fig. 4.2). Biodiesel made from palm oil may well be inferior to biodiesel when forests are cleared to grow oil palms (Danielsen et al. 2008; Fargione et al. 2008; Reijnders and Huijbregts 2008a). The same will hold for coconut biodiesel (contrasting the conclusion of Tan et al. 2004). But the opposite will hold when palms are established on abandoned agricultural land where there is currently little sequestration of C, provided that cultivation does not lead to lowering of soil carbon stocks (e.g. Germer and Sauerborn 2007). When, under market conditions, perennial grasses, short rotation woody perennials and herbaceous species such as *Miscanthus* are used in industrialized countries as lignocellulosic biomass for the production of Fischer–Tropsch diesel or ethanol, it would seem likely that there will be a large indirect effect on land use involving clearance of natural vegetation (Searchinger et al. 2008). In this case, it is doubtful whether lignocellulosic biofuels will much outperform fossil fuels regarding life cycle greenhouse gas emissions (see also the last row of Table 4.2).

### Electricity from Biowaste

There is substantial combustion of biowastes for the generation of electric power, which in principle may be used for traction. There is also substantial anaerobic conversion of biomass wastes into methane, which in turn can be used for automotive purposes. Biodiesel made from waste fats and oils (yellow grease) is currently the main biowaste-derived road transport biofuel. As pointed out in Chap. 1, there are a wide variety of processes that have been proposed to convert wastes into liquid biofuels, such as ethanol, methanol and Fischer–Tropsch liquid transport fuels, and there is also the possibility for conversion into $H_2$.

How does one evaluate the environmental impact of using biowastes to power transport? This is a tricky question. We will illustrate this with two examples, drawn from major applications of such wastes: burning to generate electricity and conversion into methane. Firstly, the allocation of emissions has a large impact on those evaluations. In this context, one may consider the example of electricity production from chicken manure in the European Union (Reijnders and Huijbregts 2005). At the time of this study, such manure had a negative price. Reijnders and Huijbregts (2005) calculated the emissions of greenhouse gases associated with electricity production using allocation on the basis of actual prices, allocation on the basis of prices but with an assumed zero price for manure and allocation on the basis of energetic outputs of chicken production. The results thereof are shown in Table 4.2.

As also pointed out in Chap. 2, so far it has been usual in life cycle assessments of biofuels from wastes to allocate on the basis of a zero price, implying that the life cycle leading to the generation of wastes has no impact on the environmental evaluation of such biofuels. A first problem with this approach is that when the current

**Table 4.2** Emission of greenhouse gases associated with 1 kWh electricity from chicken manure, using different assumptions regarding allocation (Reijnders and Huijbregts 2005)

| Assumption | Emission of greenhouse gases in g $CO_2$ equivalent per kWh; + emission − (apparent) sequestration |
|---|---|
| Allocation based on prices: negative price manure | −250 to −390 |
| Allocation based on prices: price of manure assumed to be 0 | 0 |
| Allocation based on energetic outputs poultry farming | +630 to +1,040 |

waste indeed turns out to be viable as a source of biofuel production, it will turn into a 'secondary resource' which may well have a positive price. A second complication centres around the reference to be used and the 'normal fate' of the waste used. This again may have a very large impact. For instance, a study about Dutch projects to convert manure into methane (Zwart et al. 2006) concluded that the energetic output (methane) was roughly the same as the energetic input (fossil fuels). However, Zwart et al. (2007) calculated a very low net greenhouse gas emissions for fermentation of manure, because they did not compare greenhouse gas emissions linked to methane from manure with the life cycle emissions of natural gas, but rather they compared fermentation with other ways of handling manure. Also, for the energetic application of manure, they assumed a major reduction in the emission of methane and $N_2O$ due to a much-reduced storage time for manure. So estimates about the environmental impacts of biowastes used for fuelling transport are dependent on subjective assumptions.

Zah et al. (2007) have studied emissions associated with methane production from a variety of wastes, using allocation on the basis of prices and a zero value of the waste itself. Thus, the calculation of emissions linked to methane production from wastes was restricted to the waste-to-wheel stages of the life cycle. Comparison was with natural gas. The wastes considered were: sewage sludge, biowaste, manure and manure plus co-substrates. The emission of greenhouse gases for the production of methane from these wastes was in the order of 50–80% of the fossil reference. When allocation would have been seed-to-wheel on the basis of energy or mass, the emission of greenhouse gases linked to waste-based methane production would have been much higher (cf. Table 4.2).

## 4.5 Non-greenhouse Gas Emissions

Using transport biofuels may change the emissions of non-greenhouse gases, if compared with the original (fossil) fuel. For instance, the substitution of fossil diesel by biodiesel (fatty acid ester) reduces sulphur dioxide emissions but tends to increase the emissions of nitrogen oxides ($NO_x$) from diesel, whereas the acute effects on respiratory organs do not change significantly (Ban-Weiss et al. 2007; Lin et al. 2007; Swanson et al. 2007; Szybist et al. 2007). The increase of $NO_x$ emissions caused by switching to biodiesel can be reduced by adjusting timing of the injection pump (Kegl 2008). The impact of substituting fossil diesel by biodiesel on particulate matter emissions by motorcars is apparently complex, with evidence that biodiesel substitution impacts the nanostructure of diesel soot, enhances oxidative reactivity and cytotoxicity but reduces mutagenicity (Bünger et al. 2000; Szybist et al. 2007). It appears that the overall amount of particulate matter and the number of particles that is emitted is reduced when fossil diesel is progressively replaced by biodiesel, which seems indicative of reduced risk. But the average particle size is also reduced (Kegl 2008; Keskin et al. 2008; Lapuerta et al. 2008), and smaller particle size is correlated with increased risk of a specified mass of particulate matter (Lapuerta et al. 2008). The overall effects of all these changes on human health impacts await further research (Swanson et al. 2007).

It would seem likely that, if compared with fossil gasoline, the admixture of ethanol to gasoline may be able to reduce emissions of CO and reduce ambient $O_3$ concentrations (Ahmed 2001). On the other hand, the emission of acetaldehyde is increased by such a substitution, and there may also be an increase in the atmospheric concentration of peroxylacetate nitrate (PAN) (Ahmed 2001). What the overall impact thereof on health will be awaits further research. Moreover, in practice, ethanol (or ETBE) may not substitute fossil hydrocarbons but other oxygenates of MTBE. It would seem doubtful that, as far as its impact on inhaled air is concerned, such a substitution would benefit health (Ahmed 2001).

Changes in non-greenhouse gas emissions are not confined to cars; they concern the complete life cycles. And indeed, a substantial part of the seed-to-wheel non-greenhouse gas emissions is, for instance, associated with the cropping stage. This stage is associated with the input of fertilizers ('nutrients') and pesticides. Nutrients (including conversion products thereof) may be emitted into the wider environment. Well known is the leaching of P and N nutrients into water. Leaching of these nutrients from arable soils in the US Midwest, where corn is grown to supply ethanol, is a primary contributor to the hypoxic zone in the Gulf of Mexico (Powers 2007). Hypoxic zones due to elevated levels of nutrients also occur in the East China Sea and several continental European seas, whereas continental shelves of Africa, South America and India are relatively vulnerable to increases in nutrient emissions (Diaz and Rosenberg 2008). More in general elevated concentrations of nutrients may lead to eutrophication. Eutrophication is linked with harmful algal blooms and reduced biodiversity (Granéli and Turner 2006; Ptacnik et al. 2008).

Even the cropping of *Jatropha*, which produces nuts with well-known insecticidal properties, may require substantial pesticide inputs to reduce the impact of pests (Grimm 1999; Grimm and Somarriba 1999; Carvalho et al. 2008). More generally, cropping is also associated with the use of pesticides, which may lead to ecotoxicity and toxic effects on humans. In some cases, handling of harvested materials may have a large impact on non-greenhouse gas emissions. A case in point is the burning of harvest residues of sugar cane, which serves as feedstock for the production of bioethanol. This has an adverse impact on populations living in areas where sugar cane is harvested, especially on the respiratory systems of children and the elderly (Cançado et al. 2006).

The most comprehensive study regarding transport biofuel life cycles is the work of Zah et al. (2007), who compared traditional fossil fuels with a variety of plant-based biofuels, such as rapeseed methylester, palm oil methylester, soybean methylester, methanol and ethanol from various biomass sources and countries of origin, regarding seed-to-wheel non-greenhouse gas emissions. Allocation was on the basis of prices. Zah et al. (2007) considered the life cycle emissions that may lead to oxidizing smog, eutrophication and ecotoxicity. In many cases, the emission of ecotoxic substances was found by Zah et al. (2007) to be lower for crop-based transport biofuels than for fossil fuels. However, there were also exceptions. Biodiesel from Malaysian palm oil and Brazilian soybean oil gave rise to seed-to-wheel emissions that were at least five times more ecotoxic than the fossil petrol or diesel life cycle emissions. As to eutrophication, plant-based biofuels tended to do worse than fossil transport fuels over their respective life cycles, with the exception of some wood- and grass-based products that scored rather similar to fossil transport fuels. Regarding the emission of hydrocarbons which may lead to oxidizing or photochemical smog, biofuels did often somewhat better than fossil fuels. However, soybean-based biodiesel, Malaysian oil-palm-based biodiesel and bioethanol from sugar cane in Brazil did much worse regarding their seed-to-wheel emissions of compounds that may cause oxidizing smog.

Zah et al. (2007) did not consider acidifying substances ($NO_x$, $SO_2$, $NH_3$, HCl), but other studies suggest that in this respect, biofuels often do worse than fossil fuels (Kaltschmitt et al. 1997; Sheehan et al. 2003; Reinhardt et al. 2006; Kim and Dale 2008a), when allocation is on the basis of prices. Reinhardt et al. (2006) considered a variety of processes that convert lignocellulosic biomass into transport fuels via synthesis gas. Apart from the life cycle emissions of acidifying substances, they looked at plant nutrients and compounds that may be toxic to humans, while allocating on the basis of prices, and concluded that such transport lignocellulosic biofuels did in these respects mostly worse than fossil fuels. When allocation would have been on the basis of energy content or weight of output, the emissions allocated to transport biofuels would have been lower than in the case of allocation on the basis of prices. Kim and Dale (2008a) looked at ethanol derived from US corn grain by dry milling and found that this did worse than conventional gasoline as to eutrophication and photochemical smog. In this case, allocation was done by substitution. Graebig (2006) has considered the relative environmental impacts of electricity from photovoltaics and from biogas generated by the conversion of maize. It was concluded that photovoltaics were better in all life cycle assessment categories, except eutrophication.

Zah et al. (2007) also studied waste-to-wheel emissions associated with methane production from a variety of wastes and compared these with natural gas. They found that the emission of hydrocarbons, which contribute to oxidizing smog associated with methane from wastes, was somewhat larger, and the emission of eutrophying substances much larger than in the case of natural gas. Emissions of ecotoxic substances were roughly similar or somewhat larger. The outcomes of the study of Zah et al. (2007) seem more favourable to transport biofuels made from wastes than to transport biofuels made from food crops. However, one should keep in mind that this verdict is based on the assumption that life cycle impacts up to the waste can be neglected. When wastes change into secondary resources, fetching a price, or when the allocation in life cycle assessment is based on mass or energy, differences between transport biofuels made from, for example, starch and from residues would become smaller.

## 4.6 Potential for Emission Reduction

The seed-to-wheel emissions of the transport biofuels considered here are substantial. The most important current transport biofuels (bioethanol from starch and sugar crops and biodiesel from edible vegetable oil crops) are often not a substantial improvement over current fossil transport fuels or do even worse. Thus, the question arises as to what the possibilities are for reducing this impact. Much of the impact tends to come from the production of feedstock for transport biofuel production. So the possibilities for the reduction of environmental impact associated with this stage of the transport biofuel life cycles will be considered here.

Increasing soil carbon stocks while growing feedstocks may reduce the emission of biogenic carbonaceous greenhouse gases. Reduced tillage, the use of cover crops and/or fallows, including nitrogen fixers, and the return of residues and application of other organic matter, such as manure and household composts, may contribute to such an increase in C stocks (Nandwa 2001; Bationo and Buerkert 2001; Díaz-Zorita et al. 2002; Cowie et al. 2006; Reijnders and Huijbregts 2007, 2008b). The net emissions of greenhouse gases due to changes in ecosystem carbon stocks linked to land use change following from expansion of feedstock cropping may be lowered or even reversed by growing feedstocks on soils with low aboveground carbon stocks (Germer and Sauerborn 2007). Also, increases in yield achieved at relatively low inputs of fossil fuels and improved efficiencies in converting feedstock to biofuel may reduce net greenhouse gas emissions (Gibbs et al. 2008).

In the case of $CH_4$ emission due to anaerobic conversion linked to the processing of biomass, capture of $CH_4$ and application thereof in energy generation will help (Reijnders and Huijbregts 2008a). When biofuels are burned in power plants, there is the option of $CO_2$ sequestration in abandoned gas and oil fields or aquifers, which may lead to net biogenic C sequestration (Mathews 2008). The emission of $N_2O$ may be reduced by a better N efficiency of agriculture. Precision agriculture and subsoil irrigation techniques may be conducive to a better N efficiency (Reijnders and Huijbregts 2008b). Kim and Dale (2008b) have shown that in conventional corn cropping, there is an environmentally optimal N application rate which enhances profitability to farmers. Reduction of greenhouse gas emissions linked to the production of synthetic N fertilizer may be possible by including N-fixing crops.

Thomsen and Haugaard-Nielsen (2008) have, for instance, proposed wheat undersown with clover grass for the production of biomass to be subjected to simultaneous saccharification and fermentation.

As to the emissions linked to fossil fuels, over time there have been significant reductions in the fossil fuel inputs into the production of the currently most important transport biofuel – ethanol due to efficiency gains (Hill et al. 2006; Macedo et al. 2008) – and a further significant reduction linked to efficiency gains is expected (Macedo et al. 2008).

Also, in feedstock processing, there is scope for the replacement of fossil fuels by agricultural residues, especially in the case of high-yielding crops (Reijnders and Huijbregts 2008a; Reijnders 2008). For instance, in Thailand, coal is an important fuel in the conversion of sugar cane molasses into ethanol (Nguyen et al. 2008), and coal can be replaced by residues of sugar cane or oil palm fruit processing. In the case of ethanol from sugar cane and within limitations linked to the need for maintenance of soil C stocks, the possibility exists to additionally produce electricity from bagasse (a sugar cane residue) for use elsewhere (Macedo et al. 2008).

In the case of sugar cane production, emissions of a variety of pollutants can be reduced when cane burning is replaced by mechanical harvesting (Macedo et al. 2008; Machado et al. 2008). Improving the nutrient efficiency of biofuel cropping (for instance by precision agriculture) may reduce nutrient emissions from arable land, and there is also scope for reduced pesticide emissions (e.g. Muilerman 2008).

## References


Ahmed, I (2001) Oxygenated diesel: emissions and performance characteristics of ethanol-diesel blends in CI engines. Society of Automotive Engineers 2001-01-2475. http://www.oxydiesel.com/sae2001-01-2475.pdf

Al-Kufaishi SA, Blackmore BS, Sourell H (2006) The feasibility of using variable rate water application under a central pivot irrigation system. Irrig Drain Syst 20:317–327

Almeida AC, Soares JV, Landsberg JJ, Rezende GD (2007) Growth and water balance of *Eucalyptus grandis* hybrid plantations in Brazil during a rotation for pulp production. Forest Ecol Manag 251:10–21

Ardo J, Olsson L (2003) Assessment of soil organic carbon in semi-arid Sudan using GIS and the CENTURY model. J Arid Environ 54:633–651

Banedjschafie S, Bastani S, Widmoser P, Mengel K (2008) Improvement in water use and N fertilizer efficiency by subsoil irrigation of winter wheat. Eur J Agron 28:1–7

Ban-Weiss GA, Chen JY, Buchholz BA, Dibble RW (2007) A numerical investigation into the anomalous slight $NO_x$ increase when burning biodiesel; a new (old) theory. Fuel Process Technol 88:659–667

Barnett TP, Pierce DW, Hidalgo HG, Bonfils C, Santer BD, Das T et al. (2008) Human-induced changes in the hydrology of the western United States. Science 319:1080–1083

Bationo A, Buerkert A (2001) Soil organic carbon management for sustainable land use in Sudano-Sahelian West Africa. Nutr Cycl Agroecosys 61:131–142

Bergsma G (2007) Heldergroene biomassa. [Transparently green biomass]. Stichting Natuur en Milieu, Utrecht

Berndes G (2002) Bioenergy and water – the implications of large-scale bioenergy production for water use and supply. Global Environ Chang 12:253–271

Betts R (2007) Implications of land-ecosystem-atmosphere interactions for strategies for climate change adaptation and mitigation. Tellus B 59:602–615

Betts RA, Falloon PD, Klein Goldewijk K, Ramankutty N (2007) Biophysical effects of land use change on climate: model simulations of radiative forcing and large-scale temperature change. Agr Forest Meteor 142:216–233

Börjesson P (1999) Environmental effects of energy crop cultivation in Sweden – I: identification and quantification. Biomass Bioenerg 16:137–154

Börjesson P, Mattiasson B (2008) Biogas as a resource-efficient vehicle fuel. Trends Biotechnol 26:7–13

Brady D, Pratt GC (2007) Volatile organic compound emissions from dry mill fuel ethanol production. J Air Waste Manag Assoc 57:1091–1102

Brown ME, Funk CC (2008) Food security under climate change. Science 319:580–581

Bünger J, Krahl J, Baum K, Schröder O, Müller M, Westphal G, Ruhnau P, Schulz TG, Hallier E (2000) Cytotoxic and mutagenic effects, particle size and concentration analysis of diesel engine emissions using biodiesel and petrol diesel as fuel. Arch Toxicol 74:490–498

Cançado JED, Saldiva PHN, Pereira LAA, Lara LBLS, Artaxo P, Martinelli LA, Arbex MA, Zanobetti A, Braga ALF (2006) The impact of sugar cane-burning emissions on the respiratory system of children and the elderly. Environ Health Persp 114:725–729

Carvalho CR, Clarindo WR, Praça MM, Araújo FS, Carels N (2008) Genome size, base composition and karyotype of *Jatropha curcas* L., an important biofuel plant. Plant Sci 174:613–617

Chisti Y (2007) Biodiesel from microalgae. Biotechnol Adv 25:294–306

Chisti Y (2008) Biodiesel from microalgae beats bioethanol. Trends Biotechnol 26:126–131

Cowie AL, Smith P, Johnson D (2006) Does soil carbon loss in biomass production systems negate the greenhouse benefits of bioenergy? Mitigation Adaptation Strategies Global Change 11: 979–1002

Crutzen PJ, Mosier AR, Smith KA, Winiwarter W (2007) $N_2O$ release from agro-biofuel production negates global warming reduction by replacing fossil fuels. Atmos Chem Phys 7:11191–11205

Curran MA (2007) Studying the effect of system preference by varying coproduct allocation in creating life-cycle inventory. Environ Sci Technol 41:7145–7151

Danielsen F, Beukema H, Burgess ND, Parish F, Brühl C, Donald PF, Murdiyarsno D, Phalan B, Reijnders L, Struebig M, Fitzherbert EM (2008) Biofuel plantation on forested land: double jeopardy for biodiversity and climate. Conserv Biol in press

Delire C, Behling P, Coe MT, Foley JA, Jacob R, Kutzbach J, Liu Z, Vavrus S (2001) Simulated response of the atmosphere-ocean system to deforestation in the Indonesian Archipelago. Geophys Res Lett 28:2081–2084

Delucchi MA (2005) A multi-country analysis of lifecycle emissions from transportation fuels and motor vehicles. UCD-ITS-RR-05-10. University of California at Davis, Davis

Demirbaş A (2001) Biomass resource facilities and biomass conversion processing for fuels and chemicals. Energ Convers Manage 42:1357–1378

De Resende AS, Xavier RP, de Oliveira OC, Urquiaga S, Alves BJR, Boddey RM (2006) Long-term effects of pre-harvest burning and nitrogen and vinasse applications on yield of sugar cane and soil carbon and nitrogen stocks on a plantation in Pernambuco, N.E. Brazil. Plant Soil 281:339–351

de Vries SC (2008) The bio-fuel debate and fossil energy use in palm oil production: a critique of Reijnders and Huijbregts 2007. J Clean Prod 16:1926–1927

Dias de Oliveira ME, Vaughan BE, Rykiel EJ (2005) Ethanol as fuel: energy, carbon dioxide balances, and ecological footprint. BioScience 55:593–602

Diaz RD, Rosenberg R (2008) Spreading dead zones and consequences for marine ecosystems. Science 321:926–929

Díaz-Zorita M, Duarte GA, Grove JH (2002) A review of no-till systems and soil management for sustainable crop production in the subhumid and semiarid Pampas of Argentina. Soil Till Res 65:1–18

Ding W, Cai Y, Cai Z, Yagi K, Zheng X (2007) Nitrous oxide emissions from an intensively cultivated maize–wheat rotation soil in the North China Plain. Sci Total Environ 373:501–511
Dismukes GC, Carrieri D, Bennette N, Ananyev GM, Posewitz MC (2008) Aquatic phototrophs: efficient alternatives to land-based crops for biofuels. Curr Opin Biotechnol 19:235–240
Donner SD, Kucharik CJ (2008) Corn-based ethanol production compromises goal of reducing nitrogen export by the Mississippi River. P Natl Acad Sci USA 105:4513–4518
Duc PM, Wattanavichien K (2007) Study on biogas premixed charge diesel dual fuelled engine. Energ Convers Manage 48:2286–2308
Eickhout B, van den Born GJ, Notenboom J, van Oorschot M, Ros JPM, van Vuuren DP, Westhoek HJ (2008) Local and global consequences of the EU renewable directive for biofuels. Milieu en Natuur Planbureau Bilthoven. http://www.mnp.nl
Ekvall T, Finnveden G (2001) Allocation in ISO 14041 – a critical review. J Clean Prod 9:197–208
Engelhaupt E (2007) Biofueling water problems. Environ Sci Technol 15:7593–7595
EUCAR, CONCAWE, European Commission JRC (2007) Well-to-wheels analysis of future automotive fuels and powertrains in the European context, vers 2c. http://ies.jrc.ec.europa.eu/uploads/media/WTW_Report_010307.pdf
Fargione J, Hill J, Tilman D, Polasky S, Hawthorne P (2008) Land clearing and the biofuel carbon debt. Science 319:1235–1238
Fearnside PM, Laurance WF (2004) Tropical deforestation and greenhouse gas emissions. Ecol Appl 14:982–986
Fontaine S, Barot S, Barré P, Bdioui N, Mary B, Rumpel C (2007) Stability of organic carbon in deep soil layers controlled by fresh carbon supply. Nature 450:277–280
Freibauer A, Rounsevell MDA, Smith P, Verhagen J (2004) Carbon sequestration in the agricultural soils of Europe. Geoderma 122:1–23
Fritsche UR (2007) Nachhaltige Mobilität. Zur Rolle der Biomasse [Sustainable mobility; on the role of biomass]. Öko-Institut, Darmstadt 30 November
Galloway JN, Townsend AR, Erisman JW, Bekunda M, Cai Z, Freney JR, Martinelli LA, Seitzinger SP, Sutton MA (2008) Transformation of the nitrogen cycle: recent trends, questions, and potential solutions. Science 320:889–892
Germer J, Sauerborn J (2007) Estimation of the impact of oil palm plantation establishment on greenhouse gas balance. Environ Dev Sust. doi:10.1007/s1066800690801
Gibbins J, Chalmers H (2008) Preparing for global rollout: a 'developed country first' demonstration programme for rapid CCS deployment. Energ Policy 36:501–507
Gibbs HK, Johnston M, Foley JA, Holloway T, Monfreda C, Ramankutty N, Zaks D (2008) Carbon payback times for crop-based biofuel expansion in the tropics: the effects of changing yield and technology. Environ Res Lett 3:034001
Gleick PH (2000) The world's water 2000–2001: the biennial report on freshwater resources. Island Press, Washington, DC
Gordon LJ, Peterson GD, Bennett EM (2008) Agricultural modifications of hydrological flows create ecological surprises. Trends Ecol Evol 23:211–219
Graebig M (2006) Comparative analysis of land use intensity and environmental impacts of biomass and photovoltaics. PhD Thesis. Cambridge University, Cambridge
Granéli E, Turner JT (eds) (2006) Ecology of harmful algae. Springer, Berlin
Gregorich EG, Rochette P, VandenBygaart AJ, Angers DA (2005) Greenhouse gas contributions of agricultural soils and potential mitigation practices in Eastern Canada. Soil Till Res 83:53–72
Grimm C (1999) Evaluation of damage to physic nut (*Jatropha curcas*) by true bugs. Entomol Exp Appl 92:127–136
Grimm C, Somarriba A (1999) Suitability of the physic nut (*Jatropha curcas* L.) as single host plant for the leaf-footed bug *Leptoglossus zonatus* Dallas (Het., Coreidae). J Appl Entomol 123:347–350
Grønlund A, Sveistrup TE, Søvik AK, Rasse DP, Kløve B (2006) Degradation of cultivated peat soils in northern Norway based on field scale $CO_2$, $N_2O$ and $CH_4$ emission measurements. Arch Agron Soil Sci 52:149–159

Groom MJ, Gray EM, Townsend PA (2008) Biofuels and biodiversity: principles for creating better policies for biofuel production. Conserv Biol 22:602–609

Guo J, Zhou C (2007) Greenhouse gas emissions and mitigation measures in Chinese agroecosystems. Agr Forest Meteorol 142:270–277

Gustafsson D, Lewan E, Jansson P (2004) Modeling water and heat balance of the boreal landscape – comparison of forest and arable land in Scandinavia. J Appl Meteorol 43:1750–1767

Hall DO, Rosillo-Calle F, Williams RH, Woods J (1993) Biomass for energy: supply prospects. In: Johansson TB, Kelly H, Reddy AKN, Williams RH (eds) Renewable energy sources for fuels and electricity. Island Press, Washington, DC, pp 593–651

Hammerschlag R (2006) Ethanol's energy return on investment: a survey of the literature 1990–present. Environ Sci Technol 40:1744–1750

Hill J, Nelson E, Tilman D, Polasky S, Tiffany D (2006) Environmental, economic, and energetic costs and benefits of biodiesel and ethanol fuels. P Natl Acad Sci USA 103:11206–11210

Huijbregts MAJ, Gilijamse W, Ragas AMJ, Reijnders L (2003) Evaluating uncertainty in environmental life-cycle assessment. A case study comparing two insulation options for a Dutch one-family dwelling. Environ Sci Technol 37:2600–2608

Ibragimov N, Evett SR, Esanbekov Y, Kamilov BS, Mirzaev L, Lamers JPA (2007) Water use efficiency of irrigated cotton in Uzbekistan under drip and furrow irrigation. Agr Water Manage 90:112–120

IPCC (Intergovernmental Panel on Climatic Change) (2006) Guidelines for national greenhouse gas inventories. Volume 4: Agriculture, forestry and other land use. IPCC, Geneva

Jackson RB, Jobbágy EG, Avissar R, Roy SB, Barrett DJ, Cook CW, Farley KA, le Maitre DC, McCarl BA, Murray BC (2005) Trading water for carbon with biological carbon sequestration. Science 310:1944–1947

Jantalia CP, Resck DVS, Alves BJR, Zotarelli L, Urquiaga S, Boddey RM (2007) Tillage effect on C stocks of a clayey oxisol under a soybean-based crop rotation in the Brazilian Cerrado region. Soil Till Res 95:97–109

Kaltschmitt M, Reinhardt GA, Stelzer T (1997) Life cycle analysis of biofuels under different environmental aspects. Biomass Bioenerg 12:121–134

Kegl B (2008) Effects of biodiesel on emissions of a bus diesel engine. Bioresour Technol 99:863–873

Keskin A, Gürü M, Altiparmak D, Aydin K (2008) Using of cotton oil soapstock biodiesel–diesel fuel blends as an alternative diesel fuel. Renew Energ 33:553–557

Kheshgi HS, Prince RC, Marland G (2000) The potential of biomass fuels in the context of global climate change: focus on transportation fuels. Annu Rev Energ Env 25:199–244

Kim S, Dale BE (2005) Life cycle assessment of various cropping systems utilized for producing biofuels: bioethanol and biodiesel. Biomass Bioenerg 29:426–439

Kim S, Dale BE (2008a) Life cycle assessment of fuel ethanol derived from corn grain via dry milling. Bioresour Technol 99:5250–5260

Kim S, Dale BE (2008b) Effects of nitrogen fertilizer application on greenhouse gas emissions and economics of corn production. Environ Sci Technol 42:6028–6033

Kløverpris J, Wenzel H, Nielsen PH (2008) Life cycle inventory modelling of land use induced by crop consumption. Int J Life Cycle Ass 13:13–21

Lal R (2008) Crop residues as soil amendments and feedstock for bioethanol production. Waste Manage 28:747–758

Lapuerta M, Rodríguez-Fernández J, Agudelo JR (2008) Diesel particulate emissions from used cooking oil biodiesel. Bioresour Technol 99:731–740

La Scala N Jr, Bolonhezi D, Pereira GT (2006) Short-term soil $CO_2$ emission after conventional and reduced tillage of a no-till sugar cane area in southern Brazil. Soil Till Res 91:244–248

Lemenih M, Itanna F (2004) Soil carbon stocks and turnovers in various vegetation types and arable lands along an elevation gradient in southern Ethiopia. Geoderma 123:177–188

Lin Y, Wu YG, Chang C (2007) Combustion characteristics of waste-oil produced biodiesel/diesel fuel blends. Fuel 86:1772–1780

Liu XJ, Mosier AR, Halvorson AD, Reule CA, Zhang FS (2007) Dinitrogen and $N_2O$ emissions in arable soils: effect of tillage, N source and soil moisture. Soil Biol Biochem 39:2362–2370
Liu Z, Notaro M, Kutzbach J, Liu N (2006) Assessing global vegetation-climate feedbacks from observations. J Climate 19:787–814
Lloyd SM, Ries R (2007) Characterizing, propagating, and analyzing uncertainty in life-cycle assessment: a survey of quantitative approaches. J Ind Ecol 11:161–180
Luo Y (2007) Terrestrial carbon-cycle feedback to climate warming. Annu Rev Ecol Evol S 38:683–712
Ma Z, Wood CW, Bransby DI (2000) Carbon dynamics subsequent to establishment of switchgrass. Biomass Bioenerg 18:93–104
Macedo IC, Seabra JEA, Silva JEAR (2008) Greenhouse gases emissions in the production and use of ethanol from sugarcane in Brazil: the 2005/2006 averages and a prediction for 2020. Biomass Bioenerg 32:582–595
Machado CMD, Cardoso AA, Allen AG (2008) Atmospheric emission of reactive nitrogen during biofuel ethanol production. Environ Sci Technol 42:381–385
Malça J, Freire F (2006) Renewability and life-cycle energy efficiency of bioethanol and bio-ethyl tertiary butyl ether (bioETBE): assessing the implications of allocation. Energy 31:3362–3380
Marris E (2006) Putting the carbon back: black is the new green. Nature 442:624–626
Mathews JA (2008) Carbon-negative biofuels. Energ Policy 36:940–945
Matthews RB, Pilbeam C (2005) Modelling the long-term productivity and soil fertility of maize/millet cropping systems in the mid-hills of Nepal. Agr Ecosyst Environ 111:119–139
McLaughlin SB, Adams Kszos L (2005) Development of switchgrass (*Panicum virgatum*) as a bioenergy feedstock in the United States. Biomass Bioenerg 28:515–535
McLaughlin SB, de la Torre Ugarte DG, Garten CT Jr, Lynd LR, Sanderson MA, Tolbert VR, Wolf DD (2002) High-value renewable energy from prairie grasses. Environ Sci Technol 36: 2122–2129
McPherson RA (2007) A review of vegetation–atmosphere interactions and their influences on mesoscale phenomena. Prog Phys Geog 31:261–285
Melling L, Hatano R, Goh KJ (2005a) Methane fluxes from three ecosystems in tropical peatland of Sarawak, Malaysia. Soil Biol Biochem 37:1445–1453
Melling L, Hatano R, Goh KJ (2005b) Soil $CO_2$ flux from three ecosystems in tropical peatland of Sarawak, Malaysia. Tellus B 57:1–11
Mosier A, Kroeze C, Nevison C, Oenema O, Seitzinger S, van Cleemput O (1998) Closing the global $N_2O$ budget: nitrous oxide emissions through the agricultural nitrogen cycle. Nutr Cycl Agroecosys 52:225–248
Muilerman H (2008) Integrated soybean cropping. Stichting Natuur en Milieu, Utrecht
Nandwa SM (2001) Soil organic carbon (SOC) management for sustainable productivity of cropping and agro-forestry systems in Eastern and Southern Africa. Nutr Cycl Agroecosyst 61: 143–158
Nechodom M, Schuetzle D, Ganz D, Cooper J (2008) Sustainable forests, renewable energy, and the environment. Environ Sci Technol 42:13–18
Nepstad DC, Stickler CM, Soares-Filho B, Merry F (2008) Interactions among Amazon land use, forests and climate: prospects for a near-term forest tipping point. Philos T R Soc B 363: 1737–1746
Nguyen TLT, Gheewala SH, Garivait S (2008) Full chain energy analysis of fuel ethanol from cane molasses in Thailand. Appl Energ 85:722–734
Notaro M, Liu Z, Williams JW (2006) Observed vegetation-climate feedbacks in the United States. J Climate 19:763–768
Novoa RSA, Tejeda HR (2006) Evaluation of the $N_2O$ emissions from N in plant residues as affected by environmental and management factors. Nutr Cycl Agroecosys 75:29–46
NRC (National Research Council) (2007) Water implications of biofuels production in the United States. National Academy Press, Washington, DC
Odeh NA, Cockerill TT (2008) Life cycle GHG assessment of fossil fuel power plants with carbon capture and storage. Energ Policy 36:367–380

OECD-FAO (2007) OECD-FAO Agricultural Outlook 2007–2016. OECD, Paris
Ong CK, Leakey RRB (1999) Why tree-crop interactions in agroforestry appear at odds with tree-grass interactions in tropical savannahs. Agroforest Syst 45:109–129
Ong CK, Wilson J, Deans JD, Mulayta J, Raussen T, Waja-Musukwe N (2002) Tree-crop interactions: manipulation of water use and root function. Agr Water Manag 53:171–186
Ouattara B, Ouattara K, Serpantié G, Mando A, Sédogo MP, Bationo A (2006) Intensity cultivation induced effects on soil organic carbon dynamic in the western cotton area of Burkina Faso. Nutr Cycl Agroecosys 76:331–339
Patzek TW (2004) Thermodynamics of the corn-ethanol biofuel cycle. Crit Rev Plant Sci 23: 519–567
Patzek TW (2006) A first-law thermodynamic analysis of the corn-ethanol cycle. Nat Resour Res 15:255–270
Patzek TW, Pimentel D (2005) Thermodynamics of energy production from biomass. Crit Rev Plant Sci 24:327–364
Pimentel D (2003) Ethanol fuels: energy balance, economics, and environmental impacts are negative. Nat Resour Res 12:127–134
Pimentel D, Hurd LE, Bellotti AC, Forster MJ, Oka IN, Sholes OD, Whitman RJ (1973) Food production and the energy crisis. Science 182:443–449
Pimentel D, Houser J, Preiss E, White O, Fang H, Mesnick L, Barsky T, Tariche S, Schreck J, Alpert S (1997) Water resources: agriculture, the environment, and society. An assessment of the status of water resources. BioScience 47:97–106
Postel SL, Daily GC, Ehrlich PR (1996) Human appropriation of renewable fresh water. Science 271:785–788
Powers SE (2007) Nutrient loads to surface water from row crop production. Int J Life Cycle Ass 12:399–407
Powlson DS, Goulding KWT, Willison TW, Webster CP, Hütsch BW (1997) The effect of agriculture on methane oxidation in soil. Nutr Cycl Agroecoys 49:59–70
Prateepchaikul G, Apichato T (2003) Palm oil as a fuel for agricultural diesel engines: comparative testing against diesel oil. Songklanakarin J Sci Technol 25(3). http://journeytoforever.org/biodiesel_SVO-palm.html
Pretty JN, Ball AS, Xiaoyun L, Ravindranath NH (2002) The role of sustainable agriculture and renewable resource management in reducing greenhouse-gas emissions and increasing sinks in China and India. Philos Transact A Math Phys Eng Sci 360:1741–1761
Ptacnik R, Solimini AG, Andersen T, Tamminen T, Brettum P, Lepistö L, Willén E, Rekolainen S (2008) Diversity predicts stability and resource use efficiency in natural phytoplankton communities. P Natl Acad Sci USA 105:5134–5138
Raskin P, Gleick P, Kirshen P, Pontius G, Strzepek K (1997) Water futures: assessment of long-range patterns and problems. Stockholm Environment Institute, Stockholm
Rebelo de Mira R, Kroeze C (2006) Greenhouse gas emissions from willow-based electricity: a scenario analysis for Portugal and the Netherlands. Energ Policy 34:1367–1377
Reijnders L (2008) Ethanol production from crop residues and soil organic carbon. Resour Conserv Recy 52:653–658
Reijnders L, Huijbregts MAJ (2003) Choices in calculating life cycle emissions of carbon containing gases associated with forest derived biofuels. J Clean Prod 11:527–532
Reijnders L, Huijbregts MAJ (2005) Life cycle emissions of greenhouse gases associated with burning animal wastes in countries of the European Union. J Clean Prod 13:51–56
Reijnders L, Huijbregts MAJ (2007) Life cycle greenhouse gas emissions, fossil fuel demand and solar energy conversion efficiency in European bioethanol production for automotive purposes. J Clean Prod 15:1806–1812
Reijnders L, Huijbregts MAJ (2008a) Palm oil and the emission of carbon-based greenhouse gases. J Clean Prod 16:477–482
Reijnders L, Huijbregts MAJ (2008b) Biogenic greenhouse gas emissions linked to the life cycles of biodiesel derived from European rapeseed and Brazilian soybeans. J Clean Prod 16: 1943–1948

Reijneveld A, van Wensem J, Oenema O (2009) Trends in soil organic carbon content of agricultural land in the Netherlands between 1984 and 2004. Geoderma in press

Reinhardt G, Gärtner S, Patyk A, Rettenmaier N (2006) Ökobilanzen zu BTL: Eine ökologische Einschätzung. [Biomass to liquids: an environmental assessment]. Institut für Energie- und Umweltforschung Heidelberg GmbH, Heidelberg

Renner R (2007) Rethinking biochar. Environ Sci Technol 41:5932–5933

Renouf MA, Wegener MK, Nielsen LK (2009) An environmental life cycle assessment comparing Australian sugarcane with US corn and UK sugar beet as producers of sugars for fermentation. Biomass Bioenerg in press

Righelato R, Spracklen DV (2007) Carbon mitigation by biofuels or by saving and restoring forests? Science 317:902

Rockström J, Lannerstad M, Falkenmark M (2007) Assessing the water challenge of a new green revolution in developing countries. P Natl Acad Sci USA 104:6253–6260

Romero E, Bautista J, Garcia-Martinez AM, Cremades O, Parrado J (2007) Bioconversion of corn distiller's dried grains with solubles (CDDGS) to extracellular proteases and peptones. Process Biochem 42:1492–1497

Rosell M, Lacorte S, Barcelo D (2007) Occurrence and fate of MTBE in the aquatic environment over the last decade. In: Barcelo D (ed) Fuel oxygenates: handbook of environmental chemistry part 5 R. Springer, Berlin

Royal Society (2008) Sustainable biofuels: prospects and challenges. http://royalsociety.org

Rytter RM (2005) Water use efficiency, carbon isotope discrimination and biomass production of two sugar beet varieties under well-watered and dry conditions. J Agron Crop Sci 191:426–438

Sanchez PA (1999) Delivering on the promise of agroforestry. Environ Dev Sust 1:275–284

Sarkanen KV (1976) Renewable resources for the production of fuels and chemicals. Science 191:773–776

Scharlemann JPW, Laurance WF (2008) How green are biofuels? Science 319:43–44

Scheer C, Wassmann R, Kienzler K, Ibragimov N, Eschanov R (2008) Nitrous oxide emissions from fertilized, irrigated cotton (*Gossypium hirsutum* L.) in the Aral Sea Basin, Uzbekistan: influence of nitrogen applications and irrigation practices. Soil Biol Biochem 40:290–301

Schneider N, Eugster W, Schichler B (2004) The impact of historical land-use changes on the near-surface atmospheric conditions on the Swiss Plateau. Earth Interact 8:1–27

Searchinger T, Heimlich R, Houghton RA, Dong F, Elobeid A, Fabiosa J, Tokgoz S, Hayes D, Yu T (2008) Use of U.S. croplands for biofuels increases greenhouse gases through emissions from land-use change. Science 319:1238–1240

Sheehan J, Aden A, Paustian K, Killian K, Brenner J, Walsh M, Nelson R (2003) Energy and environmental aspects of using corn stover for fuel ethanol. J Ind Ecol 7(3–4):117–146

Shrestha RK, Ladha JK, Gami SK (2006) Total and organic soil carbon in cropping systems of Nepal. Nutr Cycl Agroecosys 75:257–269

Singhal RS, Kennedy JF, Gopalakrishnan SM, Kaczmarek A, Knill CJ, Akmar PF (2008) Industrial production, processing, and utilization of sago palm-derived products. Carbohyd Polym 72: 1–20

Stape JL, Binkley D, Ryan MG (2008) Production and carbon allocation in a clonal *Eucalyptus* plantation with water and nutrient manipulations. Forest Ecol Manage 255:920–930

Su Z, Zhang J, Wu W, Cai D, Lv J, Jiang G, Huang J, Gao J, Hartmann R, Gabriels D (2007) Effects of conservation tillage practices in winter wheat water-use efficiency and crop yield on the Loess Plateau, China. Agr Water Manage 87:307–314

Sugiyama H, Fukushima Y, Hirao M, Hellweg S, Hungerbühler K (2005) Using standard statistics to consider uncertainty in industry-based life cycle inventory databases. Int J Life Cycle Ass 10:399–405

Swanson KJ, Madden MC, Ghio AJ (2007) Biodiesel exhaust: the need for health effects research. Environ Health Persp 115:496–499

Syers JK, Powlson DS, Rappaport I, Sanchez PA, Lal R, Greenland DJ, Ingram J (1997) Managing soils for long-term productivity. Philos T R Soc B 352:1011–1021

Szybist JP, Song J, Alam M, Boehman AL (2007) Biodiesel combustion, emissions and emission control. Fuel Process Technol 88:679–691

Tan KT, Lee KT, Mohamed AR, Bhatia S (2009) Palm oil: addressing issues and towards sustainable development. Renew Sust Energ Rev in press

Tan RR, Culaba AB, Purvis MRI (2004) Carbon balance implications of coconut biodiesel utilization in the Philippine automotive transport sector. Biomass Bioenerg 26:579–585

Tang H, Qiu J, Van Ranst E, Li C (2006) Estimations of soil organic carbon storage in cropland of China based on DNDC model. Geoderma 134:200–206

Taylor-Pickard J (2008) The pros and cons of feeding DDGS. Feed Mix 16(3):28–29

Thomsen MH, Haugaard-Nielsen H (2008) Sustainable bioethanol production combining biorefinery principles using combined raw materials from wheat undersown with clover-grass. J Ind Microbiol Biotechnol 35:303–311

Tilman D, Cassman KG, Matson PA, Naylor R, Polasky S (2002) Agricultural sustainability and intensive production practices. Nature 418:671–677

Tilman D, Hill J, Lehman C (2006) Carbon-negative biofuels from low-input high-diversity grassland biomass. Science 314:1598–1600

van Dijk AIJM, Keenan RJ (2007) Planted forests and water in perspective. Forest Ecol Manage 251:1–9

Vleeshouwers LM, Verhagen A (2002) Carbon emission and sequestration by agricultural land use: a model study for Europe. Glob Change Biol 8:519–530

von Blottnitz H, Curran MA (2007) A review of assessments conducted on bio-ethanol as a transportation fuel from a net energy, greenhouse gas, and environmental life cycle perspective. J Clean Prod 15:607–619

Wallace JS (2002) Increasing agricultural water use efficiency to meet future food production. Agr Ecosyst Environ 82:105–119

Weidema BP (1993) Market aspects in product life cycle inventory methodology. J Clean Prod 1:161–166

Weisser D (2007) A guide to life-cycle greenhouse gas (GHG) emissions from electric supply technologies. Energy 32:1543–1559

Weisz PB, Marshall JF (1979) High-grade fuels from biomass farming: potentials and constraints. Science 206:24–29

Wicke B, Dornburg V, Junginger M, Faaij A (2009) Different palm oil production systems for energy purposes and their greenhouse gas implications. Biomass Bioenerg in press

World Water Council (2000) World water vision. Earthscan, London

Wright L (2006) Worldwide commercial development of bioenergy with a focus on energy crop-based projects. Biomass Bioenerg 30:706–714

Yazaki Y, Mariko S, Koizumi H (2004) Carbon dynamics and budget in a *Miscanthus sinensis* grassland in Japan. Ecol Res 19:511–520

Yazdani SS, Gonzalez R (2007) Anaerobic fermentation of glycerol: a path to economic viability for the biofuels industry. Curr Opin Biotechnol 18:213–219

Zah R, Böni H, Gauch M, Hischier R, Lehman M, Wägner P (2007) Life cycle assessment of energy products: environmental impact assessment of biofuels. EMPA, St Gallen, Switzerland

Zinn YL, Lal R, Resck DVS (2005) Changes in soil organic carbon stocks under agriculture in Brazil. Soil Till Res 84:28–40

Zwart SJ, Bastiaanssen WGM (2004) Review of measured crop water productivity values for irrigated wheat, rice, cotton and maize. Agr Water Manag 69:115–133

Zwart K, Oudendag D, Ehlert P, Kuikman P (2006) Duurzaamheid en co-vergisting van dierlijke mest. Alterra, Wageningen, the Netherlands

Zwart K, Oudendag P, Kuikman P (2007) Co-digestion of animal manure and maize: is it sustainable? Alterra, Wageningen

# Chapter 5
# The Impact of Expanded Biofuel Production on Living Nature

## 5.1 Introduction

In 2007, there was a lively discussion about a plan to replace the Mabira Forest Reserve in Uganda with a sugar cane plantation for the production of the transport biofuel ethanol. This forest reserve is home to almost 300 bird species (among which is the very rare Nahan's Francolin) and supports 75 endemic species. In October 2007, the Ugandan government announced that the plan had been scrapped, because the income from conserving the Mabira Forest would dwarf the profits from bioethanol production (Williams 2007). In the case of the Mabira Forest, much of this income is derived from ecotourism (Williams 2007), with additional revenues coming from harvesting timber, making charcoal and the collection of fuelwood (Naidoo and Adamowicz 2005). There may also be other sources of income from such forests, such as the collection of food, ornamental plants and organisms that have medicinal value (Brown and Rosendo 2000; Shanley and Luz 2003; Brennan et al. 2005; Mutimukuru et al. 2006). In being a target for ecotourism and providing natural resources, living nature may be said to provide ecosystem services that have monetary value.

Decisions about clearing nature to make way for biofuel production may also turn out differently. For instance, in 2007, the replacement of the high ecological value Tanoé Swamp Forest (Ivory Coast) by an oil palm plantation for biofuel production was also started (Sielhorst et al. 2008).

Direct replacement of nature by growing biofuels is only a part of the consequences of the expansion of biofuel production. There are also indirect effects of this expansion. These follow from the relative inelasticity of the demand for food (von Braun 2007; Searchinger et al. 2008). When cropping for biofuels replaces cropping for food or feed, food or feed crops largely have to be grown somewhere else. When the expansion of cropping for biofuel production is small, it may be possible that the extra production of food and feed can be accommodated on existing agricultural soils, due to increasing productivity of agriculture. But when there is a fast and major expansion, this is not possible, and food and feed production may

L. Reijnders, M.A.J. Huibregts, *Biofuls for Road Transport*

have to expand in areas that were so far left to nature (Searchinger et al. 2008). When the expansion of biomass for biofuel production is on highly productive soils, the effect can be relatively large, as it may well be that part of the expansion of food or feed production has to take place on soils with lower productivity, which will lead to relatively large land claims.

Replacement of living nature by agriculture directly or indirectly related to the expansion of biofuel production is a significant matter in the current debate about transport biofuels. And financial considerations are important to the outcome of the conflicts between nature and (agri)culture. However, financial interests are not the only matters that are at stake. It has been argued that natural species have an intrinsic value, which requires protection against extinction. This type of argument has led to laws such as the (US) Endangered Species Act, which aims at protection of endangered species. Secondly, ecosystems provide non-monetary ecosystem services to humankind conducive to a benign biological, chemical and physical environment and to socio-cultural fulfilment (Daily 1997; Moberg and Folke 1999; Daily 2000; Batabyal et al. 2003; Díaz et al. 2006; Brauman et al. 2007; Marrs et al. 2007; Wallace 2007). The non-monetary ecosystem services include beneficial impacts on water quality and quantity, climate, soil retention, pest and disease control and pollination. A more extended list is in Table 5.1.

**Table 5.1** Non-monetary ecosystem services to mankind (Daily 1997; Díaz et al. 2006; Lelieveld et al. 2008)

| Non-monetary ecosystem service to mankind |
|---|
| – Cleansing of air and water |
| – Contribution to preservation of soil fertility and stability |
| – Regulation of water quantity and quality available to humans, crops, animal husbandry and domestic animals |
| – Pollination of plants important to humans |
| – Pest and disease control in agriculture |
| – Resistance to invasive organisms that have negative impacts |
| – Climate regulation |
| – Protection against natural hazards (floods, fires, storms) |
| – Contribution to productivity and stability of plant production important to humans |
| – Prevention of leakage of nutrients and metals from soil to surface and ground water |

Ecosystem services have been linked to biodiversity (Daily 1997, 2000), understood here as the diversity of species present in an ecosystem. There are a large number of empirical studies that underpin this link (Salonius 1981; Naeem et al. 1995; Walker 1995; van der Heijden et al. 1999; Schläpfer and Schmid 1999; Duarte 2000; Emmerson et al. 2001; Engelhardt and Ritchie 2001; Hector et al. 2001; Lyons and Schwarz 2001; Loreau et al. 2001; Tilman et al. 2001b; Duffy 2002; Emmerling et al. 2002; Symstad et al. 2003; Tilman et al. 1996; Armsworth et al. 2004; Heemsbergen et al. 2004; Reusch et al. 2005; Balvanera et al. 2006; Cardinale et al. 2006; Worm et al. 2006; Brussaard et al. 2007; Díaz et al. 2007; Fargione et al. 2007;

Turner et al. 2007; Flombaum and Sala 2008; Fornara and Tilman 2008; Ptacnik et al. 2008; Weigelt et al. 2008). From the studies done so far, it appears that there may be considerable differences between species in terms of what they contribute to ecosystem services (Cardinale et al. 2006; Jordan et al. 2006). On one hand, there are 'keystone' species that appear to have a large impact on such services. Keystone species have key functions in ecosystems. For instance, in sea grass communities, there are engineering species which, by changing the environment, facilitate the presence of species that would otherwise be absent (Duarte 2000). In arid environments, nurse plants such as *Cercidium microphyllum* and *Carnegiea gigantea* (saguaro cactus) have been identified that promote the establishment and survival of other species, including a variety of trees and shrubs (Withgott 2000; Drezner 2006, 2007). And in the US Great Basin, sagebrush serves as a nurse plant for pinyon pine (Withgott 2000).

When keystone species disappear, the loss of ecosystem services may be disproportional. On the other hand, there are species which are in one or more respects rather similar to others in what they do in ecosystems. The loss of such a species may give rise to a less-than-proportionate loss of ecosystem services. The latter is reflected in studies that suggest that halving the number of plant species, on average, leads to a reduction in primary production of about 10–20% (Tilman et al. 1996). Still, there is also the possibility that cumulative biotic changes which at first appear to have little effect may give rise to a sudden collapse of ecosystem services (Scheffer et al. 2001; Folke et al. 2004; Balmford and Bond 2005).

Such a collapse and the disproportionate effect of the loss of keystone species exemplify the possibility that the relationship between biodiversity and ecosystem services may show non-linearities (Lovelock 1988; Scheffer et al. 2001; Strange 2007). There may also be other causes for non-linearity. For instance, it has been found that there may be synergistic relations between invasive species (Grosholz 2005). And there may be interactions between losses of biodiversity and other human interventions that may give rise to non-linear effects. Malhi et al. (2008) discussed such a possibility in the context of deforestation and fire use in Amazonia. Here, there is a synergism between forest fragmentation and fire. Once burnt, a forest becomes more vulnerable to further burns and loses many primary forest species. Malhi et al. (2008) suggest that a tipping point may be reached when gasses establish in the forest understory, providing a source of fuel for repeated burns.

Monetary valuation of non-monetary ecosystem services is inherently problematic. One cannot buy them on markets. Provided that nature is there, they are freely provided. Without the ecosystem services of living nature, we would not even exist. The latter may be argued to suggest an infinite value, the former a zero value. And there have also been estimates in between (Costanza et al. 1997, 2007; Sukhdev 2008). Given the problematical character of monetary valuation of non-monetary services, the following overview considers non-monetary ecosystem services without attributing monetary values.

In Chap. 2, it was pointed out that increased yields per hectare, partly linked to intensification of agriculture, are expected to contribute to the displacement of fossil fuels by biofuels, but that large-scale displacement of fossil fuels by biofuels requires large areas of land. Current policy targets for 2020 would require the use of between 55–166 million hectares (Mha) for biofuel cropping (Renewable Fuels Agency 2008). One option is to use surplus and degraded or abandoned and fallow agricultural land for this purpose (e.g. Hoogwijk et al. 2003). An estimate suggesting that by 2050, up to 300 EJ $year^{-1}$ of liquid biofuels can be produced worldwide indeed assumes that 80% of the land area needed for that purpose will be abandoned land (de Vries et al. 2007). In this case, there will probably be a large effect on biodiversity (Huston and Marland 2003; Marland and Obersteiner 2008), and biodiversity on abandoned and fallow land is linked to ecosystem services (Börner et al. 2007; Williams et al. 2008). It is likely that abandoned and fallow land will have a lower productivity than good-quality land. As a result, when fixed amounts of transport biofuels have to be produced, as mandated under current regulations in the USA, Canada, Brazil, India and the European Union, larger areas of land will be needed for biofuel production.

It is likely that abandoned and fallow land rather often harbours substantial biodiversity (van Noordwijk 2002; Zechmeister et al. 2003; Karlowski 2006; Bowen et al. 2007; Royal Society 2008). Especially after long periods of abandonment, biodiversity may be much increased (Fournier and Planchon 1998; van Noordwijk 2002; Williams et al. 2008) and may approach biodiversity in undisturbed ecosystems. However, there are also abandoned and fallow lands with relatively low biodiversity. Examples are the *Imperata cylindrica* and *Saccharum spontaneum* dominated grasslands in formerly forested areas (Hooper et al. 2005; Germer and Sauerborn 2007). Though there are parts of such grasslands of great ecological importance (Peet et al. 1999), often they are not (MacDonald 2004). There are hundreds of mega-hectares of *Imperata cylindrica* grasslands, mainly in Africa and Asia (MacDonald 2004). In Southeastern Asia, these grasslands cover an estimated 25–35 Mha (Garrity et al. 1996; Otsamo 2000). Such grasslands are currently used for feeding livestock and thatching material (MacDonald 2004) but can also be used for biofuel production by harvesting the grasses as lignocellulosic feedstock for biofuel production or by the cultivation of, for example, short rotation woody crops that may serve as feedstock. In practice so far, the use of abandoned agricultural land for biofuel production has been very limited. For instance, expansion of Brazilian sugar cane production for the biofuel ethanol has largely been in the Cerrado region, a hotspot for biodiversity (Klink and Machado 2005; Koh 2007), and it has been suggested that further expansion may mainly take place on current pastureland (Goldemberg 2008; Goldemberg et al. 2008). Similarly, in Malaysia and Indonesia, there has been large-scale conversion of tropical forest into plantations that produce palm oil, notwithstanding the presence of large areas of degraded land in these countries (Germer and Sauerborn 2007).

There have been earlier proposals and attempts to exploit *Imperata cylindrica* grasslands for the production of wood and lignocellulosic feedstocks for the pulp and paper industry (Potter 1996; Lamb 1998; Otsamo 2000). These have met with some success (Marjokorpi and Otsamo 2006). In Southeast Asia, an estimated 2 Mha of former *Imperata cylindrica* grassland is now converted into *Acacia mangium* plantations, serving the supply of wood and lignocellulosic feedstock for the pulp and paper industry (Yamashita et al. 2008). However, attempts to convert *Imperata cylindrica* grasslands has had limited success because of limited support by local people who made use of those grasslands, e.g. for keeping livestock, and/or felt such plantations at variance with their pressing needs (Potter 1996; Otsamo 2000; Marjokorpi and Otsamo 2006), which is illustrated with a quote by one of those affected: 'The trees are healthy, but the people are sick at heart' (Potter 1996). And the production of lignocellulosic feedstock is not the only option for the conversion of *Imperata cylindrica* grasslands. It has, for instance, been suggested that, where possible, replacement of *Imperata cylindrica* grasslands by agroforests may be more in line with the needs of local populations (De Foresta and Michon 1996).

Assuming 'business as usual', a strong future expansion of transport biofuel production is expected to cause large-scale replacement of nature (Germer and Sauerborn 2007; Gurgel et al. 2007; Johansson and Azar 2007; Sivaram 2007; Christersson 2008). At a regional scale, this seems to be confirmed by studies about a future expansion of biofuel production. For instance, in studies regarding the perspectives for 'sustainable' modern biomass production in Asian countries such as China, India, Sri Lanka, Malaysia and Thailand, production of biofuels means to a considerable extent conversion of forests into plantations (Bhattacharya et al. 2003). And use of marginal lands for biofuel production in Southwestern China is doubted as much of this land is on sloping land that is prone to serious erosion (Naylor et al. 2007). In Africa, wetlands of high ecological value are increasingly considered for biofuel crops such as sugar cane and oil palm (Sielhorst et al. 2008). Moreover, the land claims associated with the expansion of transport biofuel production should be considered against the background of increasing food production.

Currently on land, about one fifth of net primary production, or somewhat more, is appropriated by humankind (Imhoff et al. 2004; Haberl et al. 2007), and about 38% of land is in agricultural use (FAO 2007). Tilman et al. (2001a) have estimated that the area needed for expansion of agriculture for food production until 2050 may be about twice as large as the area of surplus and degraded agricultural land identified by Hoogwijk et al. (2003). And the total amount of arable land where wheat, maize, oilseeds and sugar is grown for food and feed purposes is projected to grow between 2000 and 2020, while assuming substantial improvements in yield per hectare (Eickhout et al. 2008). Growing additional crops for ethanol or biodiesel production or establishing plantations of trees will, assuming business as usual, only add to the conversion of nature into 'culture', and thus to loss of habitats for living nature (Koh 2007; del Carmen Vera-Diaz et al. 2008; Searchinger et al. 2008).

Replacement is not the only impact that an expanded production of transport biofuels may have on living nature. It is likely (e.g. Goldemberg 2008) that expansion of biofuel production will partly result in intensification of agriculture, which is often associated with the increased use of inputs such as nutrients, pesticides and (irrigation) water and increased drainage (Tilman et al. 2001a; Tscharntke et al. 2005; Liira et al. 2008). This will have side effects that affect biodiversity. It is also possible to harvest nature for biomass that may serve as the basis for biofuel production. For a specified amount of biofuel, this may well affect a larger area than for a similar amount of biofuels from crops. The removal of biofuels from forests and other natural ecosystems may impact biodiversity. A case in point is the use of forestry residues, which may negatively affect a variety of species (Nordén et al. 2004; Rudolphi and Gustafsson 2005). Natural habitats may also be invaded by species planted for biofuel production (Zedler and Kercher 2004; Lavi et al. 2005; Raghu et al. 2006; Nash 2007).

Effects of replacement of nature by agriculture on biodiversity and ecosystem services will be discussed in Sect. 5.3. The effects on biodiversity and ecosystem services of cropping and harvesting practices and of invasive species used in biofuel cropping will be considered in Sects. 5.4 and 5.5, respectively. But first, in Sect. 5.2, we will go briefly into the impact of biodiversity loss on natural resources which have monetary value.

## 5.2 Loss of Biodiversity and Its Impact on Natural Resources Which Have Monetary Value

Living nature is an important supplier of natural resources which have monetary value, such as fuel, materials (e.g. rubber, aloe gel, wax, tannins, wood, thatch and broom grass), food, medicines and ornamental plants (Carr et al. 1993; Brown and Rosendo 2000; Brennan et al. 2005; Mutimukuru et al. 2006; Shackleton et al. 2007). Poor people, especially, often depend directly on the natural resources provided by living nature (Shackleton et al. 2007; Vedeld et al. 2007). An estimated more than 1.6 billion people depend for their livelihood to varying degrees on forests, and about 60 million people are fully dependent on forests (World Bank 2004). Vedeld et al. (2007) analyzed 51 case studies regarding rural dwellers from 17 developing countries in Africa, East Asia and Latin America and found that income from forests represented, on average, 22% of total income. Main contributors to income were the collection of food, fodder, fuelwood, thatch and medicine. When food prices are high, collection of wild foods has added importance for the poor (e.g. Delang 2006). In the studies reviewed by Vedeld et al. (2007), medicines from forests contributed about 7% to the income of rural dwellers. Also, natural ecosystems other than forests, such as savannahs, are important providers of natural medicines (Shackleton et al. 2007).

Currently, three-quarters of the world population depend at least partly on natural remedies (Sukhdev 2008). In China alone, 5,000 of the 30,000 recorded higher plant species are used for therapeutic purposes (Sukhdev 2008). Natural medicines have been found especially important to urban poor, for instance in countries such as South Africa and Brazil (Shanley and Luz 2003; Shackleton et al. 2007). An example of people currently affected by biodiversity loss are the city dwellers in Eastern Amazonia (Shanley and Luz 2003). Medicinal plants in this region are negatively affected by repeated cycles of forest burning and cutting and even more by the replacement of forests by biofuel crops. This, in turn, affects the availability and price of such medicinal plants, which for plants with pharmacologically demonstrated effectiveness tended to be cheaper than their counterparts from the pharmaceutical industry (Shanley and Luz 2003). There may also be a long-term effect on the availability of medicines produced by the worldwide pharmaceutical industry (Grifo et al. 1997). More than half of the medicines prescribed in the USA in 1993 contained at least one active compound 'derived from or patterned after compounds derived from biodiversity' (Grifo et al. 1997). With many species not investigated as yet as to their potential medicinal value, it may well be that the decrease of biodiversity negatively affects the future availability of new medicines.

Finally, the case of the Mabira Forest, mentioned at the beginning of this chapter, illustrated the monetary importance of tourism. Nature-oriented tourism now accounts worldwide for about 10% of international tourism expenditures and approximately 1% of total employment. It is increasingly important as a source of revenue for a wide variety of countries, accounting in some for 40–60% of all international tourists (Carr et al. 1993; Watkins 2002; Nyaupane et al. 2004; Mowforth and Munt 2005; Cochrane 2006; Shackleton et al. 2007).

## 5.3 Biodiversity and Ecosystem Function Loss Due to Replacement of Nature

Biodiversity of areas under annual crops, perennial grass crops or plantations may be different from that of the nature that they replace(d). To the extent that transport biofuel production leads to losses of natural habitats, there is apparently a consistent negative effect on species diversity (Fahrig 2003). The effect appears to be stronger at higher trophic levels (Dobson et al. 2006).

In practice, expansion of biofuel production is often associated with cutting forests, and this will probably also hold for future expansion (Gurgel et al. 2007). In such cases, land with annual biofuel crops functions in a number of other ways differently from a forest. Vegetation influences surface roughness, which in turn may alter wind, turbulence and moisture convergence, and forest and cropland may be significantly different in this respect (Notaro et al. 2006). Change of forest into cropland in semi-arid climates may increase dust emissions, which in turn may change radiative forcing (Betts 2007). The albedo of arable land, on which biofuel crops are grown, is often different from the albedo of land under natural vegetation, and actual differences may also be crop dependent (Gustafsson et al. 2004; Schneider et al. 2004; McPherson 2007). The albedo is a measure of the reflection of solar radiation by the earth's surface (including vegetation), which in turn is a determinant of net radiation. Biofuel crops tend to be shorter and have less foliage than forests, and so the surface albedo of arable land with biofuel crops tends to be higher than the albedo of a forest. The release of water to the atmosphere by evapotranspiration linked to biofuel crops may also be different from natural vegetation or other crops (Gustafsson et al. 2004; McPherson 2007). The difference in evapotranspiration may have impacts on soils and the atmosphere. An example of the former is the large-scale salinization of soils in Australia, following the replacement of native woody vegetation by crops. The change in evapotranspiration caused groundwater tables to rise, and the rising water mobilized salt (Folke et al. 2004).

As to the atmospheric effect of changed albedo, it may be noted that, besides net radiation, evapotranspiration influences local climate, including temperature (Schneider et al. 2004; Notaro et al. 2006; McPherson 2007), and there may also be an impact on precipitation (Liu et al. 2006). Moderate and local deforestation may lead locally to enhanced rainfall (Malhi et al. 2008). Also, in tropical regions, replacement of forest by annual biofuel crops may cause warming, mainly due to a decrease in evaporation and cloud cover (Betts et al. 2007). In temperate regions, the overall effect of turning forest into cropland may lead to regional cooling that may be partly offset by increased aerosol concentrations (Betts 2007).

When changes in vegetation are widespread, there may be an impact on mesoscale climate (McPherson 2007; Liu et al. 2006). For instance, a simulation study assuming a major reduction in forested area and a major increase in annual cropping in Amazonia found a substantial reduction in precipitation over Amazonia and a positive radiation forcing at the top of the atmosphere linked to increased airborne dust (Betts 2007). Another model study suggests that removal of 30–40% of the Amazonian rainforest may push much of Amazonia into a permanently drier climate (Malhi et al. 2008). Such a climate change may lead to additional damage to the current Amazonian rainforest (Nepstad et al. 2008).

Mesoscale changes in climate may have knock-on effects. For instance, mesoscale changes in Amazonia may in turn have effects on precipitation in the Northern Hemisphere and across the rest of South America (Malhi et al. 2008). A simulation regarding deforestation in Indonesia has suggested that the surrounding ocean surfaces may be warmed and that this may have a widespread impact on atmospheric circulation (Delire et al. 2001).

It has furthermore been found that cultivation of land for arable crops reduces the uptake rate of $CH_4$. In an experiment in Rothamsted, it appeared that extended (150-year) cultivation of arable land decreased $CH_4$ uptake and oxidation by 85%, if compared to that in the soil under woodland (Powlson et al. 1997). Also, the 'leakiness' of nutrients and, more in general, soluble minerals is much increased in agricultural land, if compared with soils under native forest cover (Williams and Melack 1997). Moreover, forests are better in the capture of nutrients (such as P and N) from air than annual crops (Lawrence et al. 2007). So the function of annual crops on arable land is much different from that of forests.

But how about the functioning of tree plantations, which would seem rather similar to forests? Differences in biodiversity between oil palm plantations and the forests which these plantations replaced have been analyzed by Danielsen et al. (2008). They found that total species richness of vertebrates, invertebrates and flora on oil palm plantations was impoverished. Similarly, Lindenmayer and Hobbs (2004) found reduced faunal diversity in Australian eucalypt plantations if compared with native forests. Barlow et al. (2007) studied differences between native forest and eucalypt plantations in Brazil for 15 taxonomic groups and found overall diversity reduced in plantations, with major differences in biodiversity change between the taxa. They pointed out, however, that faunal diversity can be improved by integrating elements of original biota into plantations, by modifications to plantation management (e.g. regarding harvesting and thinning) and by having extensive areas of remnant native vegetation adjacent to plantations.

The functioning of tree plantations has been found to be different from the functioning of primary forests. This has consequences for ecosystem services. Comparison of tropical tree plantations with secondary tropical forests showed, for instance, that in secondary forests, root densities and nutrient concentration in roots were higher and root penetration was deeper in forests than in plantations (Lugo 1992). This allows secondary forests to better recapture nutrients which become available by mineralization and could otherwise be lost to water and the atmosphere. In the highly productive plantations, which are important for achieving a large future biofuel supply (e.g. Hoogwijk et al. 2003), there is furthermore an intensive use of herbicides and other pesticides (Tuskan 1998; Robison and Raffa 1998) which will cause increased leakiness regarding nutrients and will limit nitrogen fixation.

Primary forests are more efficient than plantations in protecting watersheds, in reducing peak flows, soil erosion and nutrient emissions, in maintaining good water quality, in stabilizing local climate and in generating OH radicals that are important cleansing agents in the atmosphere (Hartshorn 1995; Perry 1998; Daily 2000; Monson and Holland 2001; Brauman et al. 2007; Wallace 2007). The risk of pests tends to be increased in plantations if compared with native biota, but such risk may be reduced by integrating patches of native vegetation into plantations (Lindenmayer and Hobbs 2004). Li et al. (2005) found that secondary forests may stock more long-term soil organic carbon than plantations in the wet tropics. This is relevant to the sustainability of plantations (cf. Chap. 3) and to the net emission of greenhouse gases from plantation-derived biofuels (cf. Chap. 4). Still, differences in non-monetary ecosystem services between primary forests and arable land tend to be larger than those between forests and plantations (Brauman et al. 2007).

The effect of habitat loss on non-monetary ecosystem services also has worldwide aspects. These are closely linked to the biogeochemical cycles of relatively mobile elements, of which the carbon cycle is a good example. There are several links between natural biomass and worldwide atmospheric concentrations of $CO_2$. Reduced C sequestration in ecosystems, which is often associated with expanded biofuel production (see Chap. 4), increases the atmospheric $CO_2$ concentration (Houghton 2007). It is furthermore likely that natural terrestrial biomass may give rise to a better negative feedback by fixing increasing amounts of $CO_2$ when the atmospheric concentration thereof increases than most agro-ecosystems (Cao and Woodward 1998; Luo 2007). And, overall, natural systems are better at sequestrating carbon than agricultural systems (Vitousek et al. 1997). Moreover, when land use is changed back from agriculture to nature, the functioning of secondary forests in biogeochemical cycles such as the carbon cycle may for a long time be substantially different from primary forests (Grau et al. 2003).

Losses of natural terrestrial biomass are contributing significantly to the increase in the atmospheric $CO_2$ concentration (IPCC 2001; Houghton 2007). This may also have monetary effects. An increase in atmospheric $CO_2$ concentration will (ceteris paribus) on average increase costs of production and consumption in the world economy (IPCC 2001; Stern 2007).

All in all, to the extent that transport biofuel production leads to habitat loss for living nature, this affects functions of nature that can be considered beneficial to mankind such as its contribution to a benign biological, chemical and physical environment and socio-cultural fulfilment (Vitousek et al. 1997; Wallace 2007).

## 5.4 Cropping and Harvesting Feedstocks for Biofuels

### *5.4.1 Cropping and Crop Harvesting Practices*

There are a variety of aspects of cropping and crop harvesting practices regarding terrestrial biofuels that may impact biodiversity. The use of cropping systems which include a relatively wide crop genetic diversity may allow for more services in the fields of pest control and pollination than cropping systems that have a narrow genetic base (Tilman et al. 2006; Hajjar et al. 2008). And annual cropping systems that use cover crops may well be better in soil conservation than cropping systems that do not use such crops (Jarecki and Lal 2003). There may also be differences in the non-crop biodiversity of production systems. For instance, extensively managed perennial grass crops (e.g. *Miscanthus*, switchgrass, mixtures of prairie grasses) may allow for more invertebrate diversity than intensively managed annual crops, and willow coppice plantations may benefit some bird species (Anderson and Fergusson 2006). Replacement of extensively used grasslands by arable land for biofuel cropping may negatively affect bird species that rely on such grassland habitats (e.g. Schleupner and Link 2008). Sage et al. (2006) compared bird populations in a short rotation willow coppice, used for biofuel production, and other arable crops in England and concluded that it was unlikely that the planting of willow coppice on unimproved farmland would lead to a conservation gain. However, planting willows in small blocks of different age classes and no harvest in summer would, in their view, benefit bird populations in short rotation willow coppice fields.

One likely future development to which increased transport biofuel production will contribute is intensification of agriculture (Goldemberg 2008; Searchinger et al. 2008; Sukhdev 2008). Intensified agriculture has a variety of effects on living nature. Higher production may provide more resources for a number of mammals, birds and insects. For instance, populations of bumblebees may increase in landscapes with intensive rapeseed cropping (Tscharntke et al. 2005). On the other hand, high intensities of nutrient and pesticide use tend to reduce biodiversity (Tilman et al. 2001a; Ptacnik et al. 2008). Intensification of agriculture may also decrease edge habitats such as hedges (Tscharntke et al. 2005). Intensified agriculture is furthermore often associated with lowering water tables (Tscharntke et al. 2005), and this may lead to changes in biodiversity. Also, increased irrigation, which is expected to contribute to intensified agriculture, is likely to have an impact on biodiversity by lowering the availability of water to natural species. This is exemplified by the negative impact of increased irrigation on biodiversity and ecosystem services of wetlands and rivers in the European Union, the United States, China and Australia (Gerakis and Kalburtji 1998; Gleick 2003; Castañeda and Herrero 2008; Postel 2008). On average, the net effect of intensified agriculture is a decline in biodiversity among many different taxa (Liira et al. 2008). Such a decline may have a rebound effect on crop productivity. For instance, it has been found that crop pollination from native bees may be at risk from agricultural intensification (Kremen et al. 2002).

Handling harvest residues may also matter to biodiversity. There are bird species which depend to at least some extent on harvest residues. Well known is the dependence of waterfowl on residues from harvesting in flooded rice fields (van Diepen et al. 2004), the consumption of harvest residues in the Mississippi delta by waterfowl (Gallagher et al. 2003) and the dependence of cranes on residues from corn harvesting in Northern Germany. Residue removal for biofuel production would in such cases reduce bird populations. In some cases, this may have knock-on effects. For instance, in rice-growing areas, waterfowl dependent on residues may be important in maintaining productivity as they reduce weed pressure and pests and increase N cycling (Bird et al. 2000; van Diepen et al. 2004), and they may also serve as a food source. There is little information about the impact on living nature of the cultivation of algae. Open water seaweed farming near the coast of Zanzibar has, however, been found to be associated with less sea grass and reduced abundance and biomass of macrofauna (Eklöf et al. 2005).

### *5.4.2 Harvesting Nature*

Harvesting biomass has an impact on living nature present in the location where harvesting takes place. If all net primary production were harvested, the extinction of most heterotrophic organisms would be expected. At lower levels of harvesting, food chains may still be significantly impacted (Haberl and Geissler 2000). In the case of more limited harvesting of forests, the amount of dead wood in forests is reduced, and this in turn has an impact on the many species that are dependent on dead wood (Nordén et al. 2004; Rudolphi and Gustafsson 2005). Also, long-term effects of harvesting trees have been noted on soil arthropods and the quantity of ectomycorrhizal roots in the organic horizon of forests (Mahmood et al. 1999). On the other hand, limitation of harvesting may allow for species conservation. In Swedish temperate-oak-dominated hardwood stands, a 25% harvest of understory was compatible with conservation of vascular plants, fungi, saprophytic and herbivorous beetles and mycetophilid insects (Økland et al. 2008). Management of stands on much longer than current rotations to maintain understory species, which require long periods to recover from disturbance, has been suggested as a way to limit the negative impact of harvesting on biodiversity (Halpern and Spies 1995; Kerr 1999; Ramovs and Roberts 2003). Other suggestions for forests (including 'production forests' or plantations) which are (to be) harvested have been: increasing the extent of mixed stands and improvement in vertical structure of forests through variations in stand treatments (Kerr 1999; Eriksson and Berg 2007).

All in all, harvesting of forests has been found to affect vegetation cover, animal biomass and biodiversity (Milton and Moll 1988; Halpern and Spies 1995; Carey and Johnson 1995; Chen et al. 1998). This, in turn, can entail loss of non-monetary ecosystem services that are useful to mankind (Symstad et al. 2003). For instance, in Oregon, harvesting forests increased peak flow into surface water by, on average, 30% due to the combined effect of changes in flow routing and water balance (Brauman et al. 2007). Damage to vegetation due to harvesting trees may also impact water quality. After harvesting in temperate forests, there is a transient peak in nitrate losses to surface water that may last up to 5 years (Gundersen et al. 2006). More in general, harvesting is associated with increased loss of minerals and nutrients to ground and surface water (Paré et al. 2002; Lawrence et al. 2007). Repeated harvesting in dry tropical forests may lead to depletion of nutrients to the extent that primary productivity may be negatively affected (Lawrence et al. 2007). Lowered primary productivity associated with repeated harvesting has also been noted elsewhere (Nord-Larsen 2002). Soil erosion due to harvesting trees may also be substantial (Pimentel et al. 1981). In arid environments dominated by shrubs, overharvesting may lead to loss of vegetation cover and biodiversity that may lead to desertification, including an increase in Aeolian processes such as erosion and the transportation and deposition of sand (Brown 2003; McNeely 2003).

## 5.5 Invasive Species

The possibility exists that crops that are to serve as lignocellulosic feedstocks for transport biofuel production may turn out to be invasive species. The selection for 'weedy characters' in such species, which allow for cultivation on marginal lands with relatively low inputs of nutrients, is conducive to such risk (Barney and DiTomaso 2008). The impacts on ecosystem services of invasions by species involved in biofuel production are strongly dependent on the nature of the invader and the extent of the invasion. However, effects may be considerable. One of the species considered for lignocellulosic biomass production is reed canary grass (*Phalaris arundinacea* L.). This grass species is able to invade wetlands and impact their hydrology. For the most part, the outcomes of such invasions are considered detrimental (Zedler and Kercher 2004). Stream banks may also be invaded by reed canary grass (Lavergne and Molofsky 2004). There is, furthermore, some evidence that sweet sorghum and giant reed (*Arundo donax*) are invasive in specific ecosystems in the USA (Barney and DiTomaso 2008; Royal Society 2008). *Jatropha curcas*, a source of biodiesel, is considered as invasive in South Africa and as weedy in Australia (Achten et al. 2009).

Also, the production of seaweeds for biofuel production may give rise to invasive species. A case in point is the macroalga *Kappaphycus alvarezii*. This native to the Philippines has given rise to invasions of coral reefs in Hawaiian and Indian waters as an unintended effect of cultivation (Bagla 2008).

## References

Achten WMJ, Verschot L, Franken VJ, Mathijs E, Singh VP, Aerts R, Muys B (2009) Jatropha biodiesel production and use. Biomass Bioenerg in press

Anderson GQA, Fergusson MJ (2006) Energy from biomass in the UK: sources, processes and biodiversity implications. Ibis 148:180–183

Armsworth PR, Kendall BE, Davis FW (2004) An introduction to biodiversity concepts for environmental economists. Resour Energ Econ 12:115–136

Bagla P (2008) Seaweed invader elicits angst in India. Science 320:1271

Balmford A, Bond W (2005) Trends in the state of nature and their implications for human well-being. Ecol Lett 8:1218–1234

Balvanera P, Pfisterer AB, Buchmann N, He J, Nakashizuka T, Raffaelli D, Schmid B (2006) Quantifying the evidence for biodiversity effects on ecosystem functioning and services. Ecol Lett 9:1146–1156

Barlow J, Gardner TA, Araujo IS, Ávila-Pires TC, Bonaldo AB, Costa JE et al. (2007) Quantifying the biodiversity value of tropical primary, secondary, and plantation forests. P Natl Acad Sci USA 104:18555–18560

Barney JN, DiTomaso JM (2008) Nonnative species and bioenergy: are we cultivating the next invader? BioScience 58(1):64–70

Batabyal AA, Kahn JR, O'Neill RV (2003) On the scarcity value of ecosystem services. J Environ Econ Manag 46:334–352

Betts R (2007) Implications of land-ecosystem-atmosphere interactions for strategies for climate change adaptation and mitigation. Tellus B 59:602–615

Betts RA, Falloon PD, Klein Goldewijk K, Ramankutty N (2007) Biophysical effects of land use change on climate: model simulations of radiative forcing and large-scale temperature change. Agr Forest Meteor 142:216–233

Bhattacharya SC, Abdul Salam P, Pham HL, Ravindranath NH (2003) Sustainable biomass production for energy in selected Asian countries. Biomass Bioenerg 25:471–482

Bird JA, Pettygrove GS, Eadie JM (2000) The impact of waterfowl foraging on the decomposition of rice straw: mutual benefits for rice growers and waterfowl. J Appl Ecol 37:728–741

Börner J, Mendoza A, Vosti SA (2007) Ecosystem services, agriculture, and rural poverty in the Eastern Brazilian Amazon: interrelationships and policy prescriptions. Ecol Econ 64:356–373

Bowen ME, McAlpine CA, House APN, Smith GC (2007) Regrowth forests on abandoned agricultural land: a review of their habitat values for recovering forest fauna. Biol Conserv 140:273–296

Brauman KA, Daily GC, Duarte TK, Mooney HA (2007) The nature and value of ecosystem services: an overview highlighting hydrologic services. Ann Rev Environ Resour 32:67–98

Brennan MA, Luloff AE, Finley JC (2005) Building sustainable communities in forested regions. Soc Nat Resour 18:779–789

Brown G (2003) Factors maintaining plant diversity in degraded areas of northern Kuwait. J Arid Environ 54:183–194

Brown K, Rosendo S (2000) Environmentalists, rubber tappers and empowerment: the politics and economics of extractive reserves. Dev Change 31:201–227

Brussaard L, de Ruiter PC, Brown GG (2007) Soil biodiversity for agricultural sustainability. Agr Ecosyst Environ 121:233–244

Cao M, Woodward FI (1998) Dynamic responses of terrestrial ecosystem carbon cycling to global climate change. Nature 393:249–252

Cardinale BJ, Srivastava DS, Emmett Duffy J, Wright JP, Downing AL, Sankaran M, Jouseau C (2006) Effects of biodiversity on the functioning of trophic groups and ecosystems. Nature 443:989–992

Carey AB, Johnson ML (1995) Small mammals in managed, naturally young, and old-growth forests. Ecol Appl 5:336–352

Carr TA, Pedersen HL, Ramaswamy S (1993) Rain forest entrepreneurs: cashing in on conservation. Environment 35:12–15, 33–38

Castañeda C, Herrero J (2008) Measuring the condition of saline wetlands threatened by agricultural intensification. Pedosphere 18:11–23
Chen R, Corlett RT, Hill RD (1998) The biological sustainability of biomass harvesting. Agr Ecosyst Environ 69:159–170
Christersson L (2008) Poplar plantations for paper and energy in the south of Sweden. Biomass Bioenerg 32:997–1000
Cochrane J (2006) Indonesian national parks: understanding leisure users. Ann Tourism Res 33:979–997
Costanza R, d'Arge R, de Groot R, Farber S, Grasso M, Hannon B, Limburg K, Naeem S, O'Neill RV, Paruelo J, Raskin RG, Sutton P, van den Belt M (1997) The value of the world's ecosystem services and natural capital. Nature 387:253–260
Costanza R, Fisher B, Mulder K, Liu S, Christopher T (2007) Biodiversity and ecosystem services: a multi-scale empirical study of the relation between species richness and net primary production. Ecol Econ 61:478–491
Delang CO (2006) Not just minor forest products: the economic rationale for the consumption of wild food plants by subsistence farmers. Ecol Econ 59:64–73
Daily GC (1997) Nature's services. Island Press, Washington, DC
Daily GC (2000) Management objectives for the protection of ecosystem services. Environ Sci Policy 3:333–339
Danielsen F, Beukema H, Burgess ND, Parish F, Brühl C, Donald PF, Murdiyarsno D, Phalan B, Reijnders L, Struebig M, Fitzherbert EM (2008) Biofuel plantation on forested land: double jeopardy for biodiversity and climate. Conserv Biol in press
De Foresta H, Michon G (1996) The agroforest alternative to *Imperata* grasslands: when smallholder agriculture and forestry reach sustainability. Agroforest Syst 36:105–120
Del Carmen Vera-Diaz M, Kaufmann RK, Nepstad DC, Schlesinger P (2008) An interdisciplinary model of soybean yield in the Amazon basin: the climatic, edaphic, and economic determinants. Ecol Econ 65:420–431
Delire C, Behling P, Coe MT, Foley JA, Jacob R, Kutzbach J, Liu Z, Vavrus S (2001) Simulated response of the atmosphere-ocean system to deforestation in the Indonesian Archipelago. Geophys Res Lett 28:2081–2084
de Vries BJM, van Vuuren DP, Hoogwijk MM (2007) Renewable energy sources: their global potential for the first-half of the 21st century at a global level: an integrated approach. Energ Policy 35:2590–2610
Díaz S, Fargione J, Chapin FS III, Tilman D (2006) Biodiversity loss threatens human well-being. PLoS Biol 4:1300–1305
Díaz S, Lavorel S, de Bello F, Quétier F, Grigulis K, Robson TM (2007) Incorporating plant functional diversity effects in ecosystem service assessments. P Natl Acad Sci USA 104: 20684–20689
Dobson A, Lodge D, Alder J, Cumming GS, Keymer J, McGlade J, Mooney H, Rusak JA, Sala O, Wolters V, Wall D, Winfree R, Xenopoulos MA (2006) Habitat loss, trophic collapse, and the decline of ecosystem services. Ecology 87:1915–1924
Drezner TD (2006) Plant facilitation in extreme environments: the non-random distribution of saguaro cacti (*Carnegiea gigantea*) under their nurse associates and the relationship to nurse architecture. J Arid Environ 65:46–61
Drezner TD (2007) An analysis of winter temperature and dew point under the canopy of a common Sonoran Desert nurse and the implications for positive plant interactions. J Arid Environ 69:554–568
Duarte CM (2000) Marine biodiversity and ecosystem services: an elusive link. J Exp Mar Biol Ecol 250:117–131
Duffy JE (2002) Biodiversity and ecosystem function: the consumer connection. Oikos 99:201–219
Duffy JE (2003) Biodiversity loss, trophic skew and ecosystem functioning. Ecol Lett 6:680–687
Eickhout B, van den Born GJ, Notenboom J, van Oorschot M, Ros JPM, van Vuuren DP, Westhoek HJ (2008) Local and global consequences of the EU renewable directive for biofuels. Milieu en Natuur Planbureau Bilthoven. http://www.mnp.nl

Eklöf JS, de la Torre Castro M, Adelsköld L, Jiddawi NS, Kautsky N (2005) Differences in macrofaunal and seagrass assemblages in seagrass beds with and without seaweed farms. Estuar Coast Shelf Sci 63:385–396

Emmerling C, Schloter M, Hartmann A, Kandeler E (2002) Functional diversity of soil organisms – a review of recent research activities in Germany. J Plant Nutr Soil Sci 165:408–420

Emmerson MC, Solan M, Emes C, Paterson DM, Raffaelli D (2001) Consistent patterns and the idiosyncratic effects of biodiversity in marine ecosystems. Nature 411:73–77

Engelhardt KAM, Ritchie ME (2001) Effects of macrophyte species richness on wetland ecosystem functioning and services. Nature 411:687–689

Eriksson E, Berg S (2007) Implications of environmental quality objectives on the potential of forestry to reduce net $CO_2$ emissions – a case study in central Sweden. Forestry 80:99–111

Fahrig L (2003) Effects of habitat fragmentation on biodiversity. Annu Rev Ecol Evol S 34: 487–515

FAO (Food and Agriculture Organization) (2007) http://faostat.fao.org/

Fargione J, Tilman D, Dybzinski R, Lambers JHR, Clark C, Harpole WS, Knops JMH, Reich PB, Loreau M (2007) From selection to complementarity: shifts in the causes of biodiversity–productivity relationships in a long-term biodiversity experiment. P Roy Soc B-Biol Sci 274:871–876

Flombaum P, Sala OE (2008) Higher effect of plant species diversity on productivity in natural than artificial ecosystems. P Natl Acad Sci USA 105:6087–6090

Folke C, Carpenter S, Walker B, Scheffer M, Elmqvist T, Gunderson L, Holling CS (2004) Regime shifts, resilience, and biodiversity in ecosystem management. Annu Rev Ecol Evol S 35: 557–581

Fornara DA, Tilman D (2008) Plant functional composition influences rates of soil carbon and nitrogen accumulation. J Ecol 96:314–322

Fournier A, Planchon O (1998) Link of vegetation with soil at a few metre-scale: herbaceous floristic composition and infiltrability in a Sudanian fallow-land. Acta Oecol 19:215–226

Gallagher PW, Dikeman M, Fritz J, Wailes E, Gauthier W, Shapouri H (2003) Supply and social cost estimates for biomass from crop residues in the United States. Environ Resour Econ 24:335–358

Garrity DP, Soekardi M, van Noordwijk M, de la Cruz R, Pathak PS, Gunasena HPM, van So N, Huijun G, Majid NM (1996) The *Imperata* grasslands of tropical Asia: area, distribution, and typology. Agroforest Syst 36:3–29

Gerakis A, Kalburtji K (1998) Agricultural activities affecting the functions and values of Ramsar wetland sites in Greece. Agr Ecosyst Environ 70:119–128

Germer J, Sauerborn J (2007) Estimation of the impact of oil palm plantation establishment on greenhouse gas balance. Environ Dev Sust. doi:10.1007/s1066800690801

Gessner MO, Inchausti P, Persson L, Raffaelli DG, Giller PS (2004) Biodiversity effects on ecosystem functioning: insights from aquatic systems. Oikos 104:419–422

Gleick PH (2003) Global freshwater resources: soft-path solutions for the 21st century. Science 302:1524–1528

Goldemberg J (2008) The Brazilian biofuels industry. Biotechnol Biofuels 1:6

Goldemberg J, Teixeira Coelho S, Guardabassi P (2008) The sustainability of ethanol production from sugarcane. Energ Policy 36:2086–2097

Grau HR, Aide TM, Zimmerman JK, Thomlinson JR, Helmer E, Zou X (2003) The ecological consequences of socioeconomic and land-use changes in postagriculture Puerto Rico. BioScience 53:1159–1168

Grifo F, Newman D, Fairfuield AS, Bhattacharaya B, Griupenhoff JT (1997) The origin of prescription drugs. In: Grifo F, Rosenthal J (eds) Biodiversity and human health. Island Press, Washington DC, pp 131–163

Grosholz ED (2005) Recent biological invasion may hasten invasional meltdown by accelerating historical introductions. P Natl Acad Sci USA 102:1088–1091

Gundersen P, Schmidt IK, Raulund-Rasmussen K (2006) Leaching of nitrate from temperate forests – effect of air pollution and forest management. Environ Rev 14:1–57

Gurgel A, Reilly JM, Paltsev S (2007) Potential land use implications of a global biofuels industry. J Agric Food Ind Organ 5(2):article 9. http://www.bepress.com/jafio/vol5/iss2/art9/

Gustafsson D, Lewan E, Jansson P (2004) Modeling water and heat balance of the boreal landscape – comparison of forest and arable land in Scandinavia. J Appl Meteorol 43:1750–1767

Haberl H, Geissler S (2000) Cascade utilization of biomass: strategies for a more efficient use of a scarce resource. Ecol Eng 16:S111–S121

Haberl H, Erb KH, Krausmann F, Gaube V, Bondeau A, Plutzar C, Gingrich S, Lucht W, Fischer-Kowalski M (2007) Quantifying and mapping the human appropriation of net primary production in earth's terrestrial ecosystems. P Natl Acad Sci USA 104:12942–12947

Hajjar R, Jarvis DI, Gemmill-Herren B (2008) The utility of crop genetic diversity in maintaining ecosystem services. Agr Ecosyst Environ 123:261–270

Halpern CB, Spies TA (1995) Plant species diversity in natural and managed forests of the Pacific Northwest. Ecol Appl 5:913–934

Hamilton C (1997) The sustainability of logging in Indonesia's tropical forests: a dynamic input-output analysis. Ecol Econ 21:183–195

Hartshorn GS (1995) Ecological basis for sustainable development in tropical forests. Annu Rev Ecol Syst 26:155–175

Hector A, Dobson K, Minns A, Bazeley-White E, Lawton JH (2001) Community diversity and invasion resistance: an experimental test in a grassland ecosystem and a review of comparable studies. Ecol Res 16:819–831

Heemsbergen DA, Berg MP, Loreau M, van Hal JR, Faber JH, Verhoef HA (2004) Biodiversity effects on soil processes explained by interspecific functional dissimilarity. Science 306: 1019–1020

Hoogwijk M, Faaij A, van den Broek R, Berndes G, Gielen D, Turkenburg W (2003) Exploration of the ranges of the global potential of biomass for energy. Biomass Bioenerg 25:119–133

Hooper E, Legendre P, Condit R (2005) Barriers to forest regeneration of deforested and abandoned land in Panama. J Appl Ecol 42:1165–1174

Houghton RA (2007) Balancing the global carbon budget. Annu Rev Earth Pl Sci 35:315–347

Huston MA, Marland G (2003) Carbon management and biodiversity. J Environ Manag 67:77–86

Imhoff ML, Bounoua L, Ricketts T, Loucks C, Harriss R, Lawrence WT (2004) Global patterns in human consumption of net primary production. Nature 429:870–873

IPCC (2001) Climate change. Cambridge University Press, Cambridge

Jarecki MK, Lal R (2003) Crop management for soil carbon sequestration. Crit Rev Plant Sci 22:471–502

Johansson DJA, Azar C (2007) A scenario based analysis of land competition between food and bioenergy production in the US. Climatic Change 82:267–291

Jordan F, Liu W, Davis AJ (2006) Topological keystone species: measures of positional importance in food webs. Oikos 112:535–546

Karlowski U (2006) Afromontane old-field vegetation: secondary succession and the return of indigenous species. Afr J Ecol 44:264–272

Kerr G (1999) The use of silvicultural systems to enhance the biological diversity of plantation forests in Britain. Forestry 72:191–205

Klink CA, Machado RB (2005) Conservation of the Brazilian Cerrado. Conserv Biol 19:707–713

Koh LP (2007) Potential habitat and biodiversity losses from intensified biodiesel feedstock production. Conserv Biol 21:1373–1375

Kremen C, Williams NM, Thorp RW (2002) Crop pollination from native bees at risk from agricultural intensification. P Natl Acad Sci USA 99:16812–16816

Lamb D (1998) Large-scale ecological restoration of degraded tropical forest lands: the potential role of timber plantations. Restor Ecol 6:271–279

Lavergne S, Molofsky J (2004) Reed canary grass (*Phalaris arundinacea*) as a biological model in the study of plant invasions. Crit Rev Plant Sci 23:415–429

Lavi A, Perevolotsky A, Kigel J, Noy-Meir I (2005) Invasion of *Pinus halepensis* from plantations into adjacent natural habitats. Appl Veg Sci 8:85–92

Lawrence D, D'Odorico P, Diekmann L, DeLonge M, Das R, Eaton J (2007) Ecological feedbacks following deforestation create the potential for a catastrophic ecosystem shift in tropical dry forest. P Natl Acad Sci USA 104:20696–20701

Lelieveld J, Butler TM, Crowley JN, Dillon TJ, Fischer H, Ganzeveld L, Harder H, Lawrence MG, Martinez M, Taraborrelli D, Williams J (2008) Atmospheric oxidation capacity sustained by a tropical forest. Nature 452:737–740

Li Y, Xu M, Zou X, Shi P, Zhang Y (2005) Comparing soil organic carbon dynamics in plantation and secondary forest in the wet tropics in Puerto Rico. Glob Change Biol 11:239–248

Liira J, Schmidt T, Aavik T, Arens P, Augenstein I, Bailey D, Billeter R, Bukáček R, Burel F et al. (2008) Plant functional group competition and large-scale species richness in European agricultural landscapes. J Veg Sci 19:3–14

Lindenmayer DB, Hobbs RJ (2004) Fauna conservation in Australian plantation forests – a review. Biol Conserv 119:151–168

Liu Z, Notaro M, Kutzbach J, Liu N (2006) Assessing global vegetation–climate feedbacks from observations. J Climate 19:787–814

Loreau M, Naeem S, Inchausti P, Bengtsson J, Grime JP, Hector A, Hooper DU, Huston MA, Raffaelli D, Schmid B, Tilman D, Wardle DA (2001) Biodiversity and ecosystem functioning: current knowledge and future challenges. Science 294:804–808

Lovelock J (1988) Ages of Gaia: a biography of our living earth. Oxford University Press, Oxford

Lugo AE (1992) Comparison of tropical tree plantations with secondary forests of similar age. Ecol Monogr 62:1–41

Luo Y (2007) Terrestrial carbon-cycle feedback to climate warming. Annu Rev Ecol Evol S 38: 683–712

Lyons KG, Schwartz MW (2001) Rare species loss alters ecosystem function – invasion resistance. Ecol Lett 4:358–365

MacDonald GE (2004) Cogongrass (*Imperata cylindrica*) – biology, ecology and management. Crit Rev Plant Sci 23:367–380

Mahmood S, Finlay RD, Erland S (1999) Effects of repeated harvesting of forest residues on the ectomycorrhizal community in a Swedish spruce forest. New Phytol 142:577–585

Malhi Y, Roberts JT, Betts RA, Killeen TJ, Li W, Nobre CA (2008) Climate change, deforestation, and the fate of the Amazon. Science 319:169–172

Marjokorpi A, Otsamo R (2006) Prioritization of target areas for rehabilitation: a case study from West Kalimantan, Indonesia. Restor Ecol 14:662–673

Marland G, Obersteiner M (2008) Large-scale biomass for energy, with considerations and caution: an editorial comment. Climatic Change 87:335–342

Marrs RH, Galtress K, Tong C, Cox ES, Blackbird SJ, Heyes TJ, Pakeman RJ, Le Duc MG (2007) Competing conservation goals, biodiversity or ecosystem services: element losses and species recruitment in a managed moorland-bracken model system. J Environ Manag 85:1034–1047

Martens P, Rotmans J, de Groot D (2003) Biodiversity: luxury or necessity? Glob Environ Change 13:75–81

McNeely JA (2003) Biodiversity in arid regions: values and perceptions. J Arid Environ 54:61–70

McPherson RA (2007) A review of vegetation–atmosphere interactions and their influences on mesoscale phenomena. Prog Phys Geog 31:261–285

Mikola J (1998) Effects of microbivore species composition and basal resource enrichment on on trophic-level biomasses in an experimental microbial-based soil food web. Oecologia 117: 396–403

Milton SJ, Moll EJ (1988) Effects of harvesting on frond production of *Rumohra adiantiformis* in South Africa. J Appl Ecol 25:725–743

Moberg F, Folke C (1999) Ecological goods and services of coral reef ecosystems. Ecol Econ 29:215–233

Monson RK, Holland EA (2001) Biospheric trace gas fluxes and their control over tropospheric chemistry. Annu Rev Ecol Syst 32:547–576

Mowforth M, Munt I (2005) Tourism and sustainability. Routledge, London

Mutimukuru T, Kozanayi W, Nyirenda R (2006) Catalyzing collaborative monitoring processes in joint forest management situations: the Mafungautsi Forest case, Zimbabwe. Soc Nat Resour 19:209–224

Naeem S, Thompson LJ, Lawler SP, Lawton JH, Woodfin RM (1995) Empirical evidence that declining species diversity may alter the performance of terrestrial ecosystems. Philos T R Soc B 347:249–262

Naidoo R, Adamowicz WL (2005) Economic benefits of biodiversity exceed costs of conservation at an African rainforest reserve. P Natl Acad Sci USA 102:16712–16716

Nash S (2007) Decrypting biofuel scenarios. BioScience 57:472–477

Naylor RL, Liska AJ, Burke MB, Falcon WP, Gaskell JC, Rozelle SD, Cassman KG (2007) The ripple effect: biofuels, food security, and the environment. Environment 49(9):30–43

Nepstad DC, Stickler CM, Soares-Filho B, Merry F (2008) Interactions among Amazon land use, forests and climate: prospects for a near-term forest tipping point. Philos T R Soc B 363: 1737–1746

Nord-Larsen T (2002) Stand and site productivity response following whole-tree harvesting in early thinnings of Norway spruce. Biomass Bioenerg 23:1–12

Nordén B, Götmark F, Tönnberg M, Ryberg M (2004) Dead wood in semi-natural temperate broadleaved woodland: contribution of coarse and fine dead wood, attached dead wood and stumps. Forest Ecol Manag 194:235–248

Notaro M, Liu Z, Williams JW (2006) Observed vegetation-climate feedbacks in the United States. J Climate 19:763–768

Nugteren HW, Janssen-Jurkovicova M, Scarlett B (2001) Improvement of environmental quality of coal fly ash by applying forced leaching. Fuel 80:873–877

Nyaupane GP, Morais DB, Graefe AR (2004) Nature tourism constraints: a cross-activity comparison. Ann Tourism Res 31:540–555

Økland B, Götmark F, Nordén B (2008) Oak woodland restoration: testing the effects on biodiversity of mycetophilids in southern Sweden. Biodivers Conserv. doi:10.1007/s1053100893254

Otsamo R (2000) Early development of three planted indigenous tree species and natural understorey vegetation in artificial gaps in an *Acacia mangium* stand on an *Imperata cylindrica* grassland site in South Kalimantan, Indonesia. New Forests 19:51–68

Paine RT (2002) Trophic control of production in a rocky intertidal community. Science 296: 736–739

Paré D, Rochon P, Brais S (2002) Assessing the geochemical balance of managed boreal forests. Ecol Indic 1:293–311

Peet NB, Watkinson AR, Bell DJ, Kattel BJ (1999) Plant diversity in the threatened sub-tropical grasslands of Nepal. Biol Conserv 88:193–206

Perry DA (1998) The scientific basis of forestry. Annu Rev Ecol Syst 29:435–466

Petchey OL, Downing AL, Mittelbach GG, Persson L, Steiner CF, Warren PH, Woodward G (2004) Species loss and the structure and functioning of multitrophic aquatic systems. Oikos 104: 467–478

Pimentel D, Moran MA, Fast S, Weber G, Bukantis R, Balliett L, Boveng P, Cleveland C, Hindman S, Young M (1981) Biomass energy from crop and forest residues. Science 212:1110–1115

Postel SL (2008) The forgotten infrastructure: safeguarding freshwater ecosystems. J Int Aff 61(2):75–90

Potter LM (1996) The dynamics of *Imperata*: historical overview and current farmer perspectives, with special reference to South Kalimantan, Indonesia. Agroforest Syst 36:31–51

Powlson DS, Goulding KWT, Willison TW, Webster CP, Hütsch BW (1997) The effect of agriculture on methane oxidation in soil. Nutr Cycl Agroecoys 49:59–70

Ptacnik R, Solimini AG, Andersen T, Tamminen T, Brettum P, Lepistö L, Willén E, Rekolainen S (2008) Diversity predicts stability and resource use efficiency in natural phytoplankton communities. P Natl Acad Sci USA 105:5134–5138

Raghu S, Anderson RC, Daehler CC, Davis AS, Wiedenmann RN, Simberloff D, Mack RN (2006) Adding biofuels to the invasive species fire? Science 313:1742

Ramovs BV, Roberts MR (2003) Understory vegetation and environment responses to tillage, forest harvesting, and conifer plantation development. Ecol Appl 13:1682–1700

Reddy MS, Venkataraman C (2002) Inventory of aerosol and sulphur dioxide emissions from India. Part II – biomass combustion. Atmos Environ 36:699–712

Reid H, Swiderska K (2008) Biodiversity, climate change and poverty: exploring the links. IIED Briefing Papers February. http://www.iied.org/pubs/display.php?o=17034IIED

Renewable Fuels Agency (2008) The Gallagher review of the indirect effects of biofuels. Renewable Fuels Agency, St Leonards-on-Sea (East Sussex, UK)

Reusch TBH, Ehlers A, Hâmmerli A, Worm B (2005) Ecosystem recovery after climatic extremes enhanced by genotypic diversity. P Natl Acad Sci USA 102:2826–2831

Robison DJ, Raffa KF (1998) Productivity, drought tolerance and pest status of hybrid Populus: tree improvement and silvicultural implications. Biomass Bioenerg 14:1–20

Royal Society (2008) Sustainable biofuels: prospects and challenges. http://royalsociety.org

Rudolphi J, Gustafsson L (2005) Effects of forest-fuel harvesting on the amount of deadwood on clear-cuts. Scand J Forest Res 20:235–242

Sage R, Cunningham M, Boatman N (2006) Birds in willow short-rotation coppice compared to other arable crops in central England and a review of bird census data from energy crops in the UK. Ibis 146:184–197

Salonius PO (1981) Metabolic capabilities of forest soil microbial populations with reduced species diversity. Soil Biol Biochem 13:1–10

Scheffer M, Carpenter S, Foley JA, Folke C, Walker B (2001) Catastrophic shifts in ecosystems. Nature 413:591–596

Schläpfer F, Schmid B (1999) Ecosystem effects of biodiversity: a classification of hypotheses and exploration of empirical results. Ecol Appl 9:893–912

Schleupner C, Link PM (2008) Potential impacts on important bird habitats in Eiderstedt (Schleswig-Holstein) caused by agricultural land use changes. Appl Geogr 28:237–247

Schneider N, Eugster W, Schichler B (2004) The impact of historical land-use changes on the near-surface atmospheric conditions on the Swiss Plateau. Earth Interact 8:1–27

Searchinger T, Heimlich R, Houghton RA, Dong F, Elobeid A, Fabiosa J, Tokgoz S, Hayes D, Yu T (2008) Use of U.S. croplands for biofuels increases greenhouse gases through emissions from land-use change. Science 319:1238–1240

Shackleton CM, Schackleton SE, Buiten E, Bird N (2007) The importance of dry woodlands and forests in rural livelihoods and poverty alleviation in South Africa. Forest Policy Econ 9: 558–577

Shanley P, Luz L (2003) The impacts of forest degradation on medicinal plant use and implications for health care in Eastern Amazonia. BioScience 53:573–584

Sielhorst S, Molenaar JW, Offermans D (2008) Biofuels in Africa. Wetlands International, Wageningen

Sivaram S (2007) Materials and energy derived from carbohydrates: opportunities, challenges and sustainability assessment. Chemical Business May:31–36

Stern, N (2007) The economics of climate change: the Stern review. Cambridge University, Cambridge

Strange CJ (2007) Facing the brink without crossing it. BioScience 57:920–926

Sukhdev P (2008) The economics of ecosystems & biodiversity. European Communities, Brussels

Symstad AJ, Chapin FS III, Wall DH, Gross KL, Huenneke LF, Mittelbach GG, Peters DPC, Tilman D (2003) Long-term and large-scale perspectives on the relationship between biodiversity and ecosystem functioning. BioScience 53:89–98

Tilman D, Wedin D, Knops J (1996) Productivity and sustainability influenced by biodiversity in grassland ecosystems. Nature 379:718–720

Tilman D, Fargione J, Wolff B, D'Antonio C, Dobson A, Howarth R, Schindler D, Schlesinger WH, Simberloff D, Swackhamer D (2001a) Forecasting agriculturally driven global environmental change. Science 292:281–284

Tilman D, Reich PB, Knops J, Wedin D, Mielke T, Lehman C (2001b) Diversity and productivity in a long-term grassland experiment. Science 294:843–845

Tilman D, Hill J, Lehman C (2006) Carbon-negative biofuels from low-input high-diversity grassland biomass. Science 314:1598–1600

Tscharntke T, Klein AM, Kruess A, Steffan-Dewenter I, Thies C (2005) Landscape perspectives on agricultural intensification and biodiversity – ecosystem service management. Ecol Lett 8: 857–874

Turner WR, Brandon K, Brooks TM, Costanza R, da Fonseca GAB, Portela R (2007) Global conservation of biodiversity and ecosystem services. BioScience 57:868–873

Tuskan GA (1998) Short-rotation woody crop supply systems in the United States: what do we know and what do we need to know? Biomass Bioenerg 14:307–315

van der Heijden MGA, Klironomos JN, Ursic M, Moutoglis P, Streitwolf-Engel R, Boller T, Wiemken A, Sanders IR (1999) "Sampling effect", a problem in biodiversity manipulation? A reply to David A. Wardle. Oikos 87:408–410

van Diepen LTA, van Groenigen JW, van Kessel C (2004) Isotopic evidence for changes in residue decomposition and N-cycling in winter flooded rice fields by foraging waterfowl. Agr Ecosyst Environ 102:41–47

van Noordwijk M (2002) Scaling trade-offs between crop productivity, carbon stocks and biodiversity in shifting cultivation landscape mosaics: the FALLOW model. Ecol Model 149:113–126

Vedeld P, Angelsen A, Bojo J, Sjaastad E, Berg GK (2007) Forest environmental incomes and the rural poor. Forest Policy Econ 9:869–879

Vitousek PM, Mooney HA, Lubchenco J, Melillo JM (1997) Human domination of Earth's ecosystem. Science 277:494–499

von Braun J (2007) The world food situation. International Food Policy Research Institute, Washington DC

Walker B (1995) Conserving biological diversity through ecosystem resilience. Conserv Biol 9:747–752

Wallace KJ (2007) Classification of ecosystem services: problems and solutions. Biol Conserv 139:235–246

Wardle DA, Bonner KI, Barker GM, Yeates GW, Nicholson KS, Bardgett RD, Watson RN, Ghani A (1999) Plant removals in perennial grassland: vegetation dynamics, decomposers, soil biodiversity, and ecosystem properties. Ecol Monogr 69:535–568

Watkins JB (2002) The evolution of ecotourism in East Africa: from idea to industry. International Institute for Environment and Development, Nairobi

Weigelt A, Schumacher J, Roscher C, Schmid B (2008) Does biodiversity increase spatial stability in plant community biomass? Ecol Lett 11:338–347

Williams MR, Melack JM (1997) Solute export from forested and partially deforested catchments in the central Amazon. Biogeochemistry 38:67–102

Williams M, Ryan CM, Rees RM, Sambane E, Fernando J, Grace J (2008) Carbon sequestration and biodiversity of re-growing miombo woodlands in Mozambique. Forest Ecol Manag 254:145–155

Williams N (2007) Biofuel backfire fears. Curr Biol 17:R983–R984

Withgott J (2000) Botanical nursing: from deserts to shorelines, nurse effects are receiving renewed attention. BioScience 50:479–484

World Bank (2004) Sustaining forests: a development strategy. World Bank, Washington, DC

Worm B, Barbier EB, Beaumont N, Duffy JE, Folke C, Halpern BS, Jackson JBC, Lotze HK, Micheli F, Palumbi SR, Sala E, Selkoe KA, Stachowicz JJ, Watson R (2006) Impacts of biodiversity loss on ocean ecosystem services. Science 314:787–790

Yamashita N, Ohta S, Hardjono A (2008) Soil changes induced by *Acacia mangium* plantation establishment: comparison with secondary forest and *Imperata cylindrical* grassland soils in South Sumatra, Indonesia. Forest Ecol Manag 254:362–370

Zechmeister HG, Tribsch A, Moser D, Peterseil J, Wrbka T (2003) Biodiversity 'hot spots' for bryophytes in landscapes dominated by agriculture in Austria. Agr Ecosyst Environ 94: 159–167

Zedler JB, Kercher S (2004) Causes and consequences of invasive plants in wetlands: opportunities, opportunists and outcomes. Crit Rev Plant Sci 23:431–452

# Chapter 6
# Frequently Asked Questions in the Transport Biofuel Debate

## 6.1 Introduction

As pointed out in Chap. 1, there is a lively debate about the future of transport biofuels. In this debate, many matters have been raised. Some of these refer to the question of whether biofuels do what they have been promised to do, e.g. increase energy security and tackle climate change. Others refer to the side effects of transport biofuel production, for instance on biodiversity, natural resources such as water and food prices. And there are questions that relate to what governments should do about transport biofuels and how to proceed with specific feedstocks: should they be processed in biorefineries, or rather be converted into one biofuel?

In this chapter, we try to answer several of the questions that have frequently been raised in the transport biofuel debate. In doing so, we will draw on the previous chapters.

The frequently asked questions that we try to answer in this chapter are:

- Should the focus be on one-fuel output or on multi-output biorefineries?
- Can transport biofuels significantly contribute to energy security?
- What is the effect of transport biofuel production on food security and food prices?
- Is expanding biofuel production a good way to tackle climate change?
- What is the effect of biofuel production on nature conservation?
- How to use natural resources in biofuel production in a sustainable way?
- What government policy should one aim at for biofuels?

## 6.2 A Focus on One Transport Biofuel Output or on Biorefineries?

A trend in the production of ethanol is increased interest in the combined conversion of starch, cellulose and hemicellulose into ethanol. In this way, the traditional production of two outputs (ethanol and dried distillers grains with or without solubles – to be used in, for example, animal feed) is replaced by one output: ethanol (Linde et al. 2008). Additionally, efforts are under way to eliminate a second co-product from ethanol production: glycerol (Bideaux et al. 2006). Similarly, in butanol production, there is currently much effort focused on getting rid of the by-products acetone and ethanol (Antoni et al. 2007; Dürre 2008). Also, anaerobic conversion by mixtures of micro-organisms converts a wide variety of organic substances to one fuel: methane.

On the other hand, there is also a trend to widening the variety of outputs of production processes generating transport biofuels. At the factory level, the analogy to petrochemical refineries has given rise to the concept of a 'biorefinery' that produces a variety of products from biomass feedstocks (Kamm and Kamm 2004; Arifeen 2007; Hayes 2008). In producing more than one product in the context of fermentative ethanol production, a variety of technologies may be used, including several biotechnologies and chemical synthesis technologies, the latter starting from 'platform chemicals' such as levulinic acid (Hayes 2008; Huang et al. 2008). When synthesis gas or hydrocarbons are produced from biomass, one might envisage the development of refineries with synthetic technologies which are now commonly applied in the petrochemical industry (e.g. Chew and Bhatia 2008; Rowlands et al. 2008). Biorefinery concepts including the production of monomers for current bulk chemicals such as eth(yl)ene and caprolactam have been proposed, too (Kamm and Kamm 2004). Also, biorefineries based on hydrocarbons containing oxygen have been suggested starting from pyrolysis oil (Hayes 2008).

In line with the biorefinery concepts focusing on the use of biotechnology and separation technologies, there is, for instance, an operational factory that converts corn into the biofuel ethanol and also produces citric acid, lactic acid, amino acids and enzymes. And a wheat biorefinery has been proposed generating, besides ethanol, ferulic acid, arabinoxylan, amino acids and gluten (Arifeen et al. 2007).

There is probably a place for both the biorefinery and single output approaches. And both may have significant implications elsewhere in the economy. Eliminating current by-products of biofuel production which are used as animal feed, such as dried distillers grains, will probably have an upward effect on other types of animal feed production (Searchinger et al. 2008).

As pointed out in Sect. 1.7, if transport biofuels are going to replace current fossil fuels on a large scale by multi-output types of production, markets for by-products may be easily flooded, leading, among other things, to major price reductions for such by-products. This has already happened in the case of glycerol, a by-product of biodiesel production (Yazdani and Gonzalez 2007). In the summer of 2007, a glut of dried distillers grains with solubles, a co-product of bioethanol production, in the USA led to relatively low prices paid for its use as an ingredient in animal feed (Tyner 2008). And flooding markets by biorefineries may also happen in marketing co-products such as xylitol, xylo-oligosaccharides and lignin (Kadam et al. 2008).

## 6.3 Can Transport Biofuels Significantly Contribute to Energy Security?

As pointed out in Chap. 2, current transport biofuels often have a cumulative seed-to-wheel fossil energy demand that tends to be smaller than the fossil fuels that they replace. The difference with fossil transport fuels is variable. The difference is probably about zero for $CH_4$ from manure in NW Europe; is as yet unlikely to be positive for current algal biofuels; is smaller than 40% for ethanol from European wheat, biodiesel from European rapeseed or ethanol from US corn and is relatively large for palm oil biodiesel and ethanol from sugar cane, especially when processing is powered by agricultural residues. When the latter applies, the difference may, for example, become greater than 90% for ethanol from sugar cane (Macedo et al. 2008). The overall solar energy conversion efficiency of biofuels suitable for use in internal combustion engines is probably around 0.2% or lower, and expected yield increases per hectare are in the order of roughly 1% per year (cf. Chap. 2). This means that to displace substantial amounts of fossil fuels, land requirements for such transport biofuels are large, as also noted by Dukes (2003).

The contribution that biofuels can make to the national energy security of a country depends on the magnitude of fuel demand and the land area available for supplying biofuel feedstocks. For a country such as Brazil, ethanol from sugar cane can make a significant contribution to national energy security. In the USA, the contribution of biofuels to national energy security is likely to be much smaller. The reasons for this are that, if compared with Brazil, per capita demand for transport fuels is larger, per capita land availability is lower and feedstocks such as corn and canola are less efficient converters of solar energy into biofuel than sugar cane. Eaves and Eaves (2007) have a point when they argue that devoting 100% of US corn to ethanol, while correcting for fossil fuel inputs, would displace 3.5% of gasoline consumption, 'only slightly more than the displacement that would follow from properly inflated tires'. In fact, they may even have been too optimistic, because the actual US policy has been using a federal excise tax, making mixtures of conventional gasoline and ethanol cheaper than conventional gasoline, which has an upward effect on overall transport fuel use (Vedenov and Wetzstein 2008). Diverting all 2007 US soybean cultivation to biodiesel production would cover approximately 2% of US diesel demand, when corrected for fossil fuel inputs and assuming no effect on fuel prices (Bagajewicz et al. 2007; Reijnders and Huijbregts 2008b).

Larger displacements of fossil fuels while using the same area of land can be achieved when biomass is burned in power stations and used for electric traction, as the seed-to-wheel overall solar conversion efficiency thereof is higher than in the case of transport biofuels such as biodiesel and bioethanol, as indicated in Chap. 2. However, as explained in Sect. 1.6, such a strategy is dependent on a major change in social acceptance of plug-in vehicles. The potential for energy security through national transport biofuel supply is low for industrialized countries with high population densities, such as Japan and the Low Countries in Europe. For instance, in the Netherlands, a 20% target for the share of transport biofuels in current transport fuel consumption would require an area of arable land that is roughly four to five times the size of current agricultural land in that country when ethanol from starch and sugar and biodiesel from vegetable oil are used and when a correction is made for the cumulative fossil fuel inputs in the biofuel lifecycles.

Not only the land area available, but also other factors may limit the extent to which countries may rely on domestically produced biofuel feedstocks for energy security. Climatic change may well have a negative impact on agricultural yields in the developing world (Jepma 2008). As pointed out in Chap. 3, water requirements for producing substantial amounts of biofuel feedstocks are large, and currently, structural water shortages affect about 300–400 million people mainly in Africa and Asia in a band from China to North Africa. Large additional water requirements follow from expected population growth and changes in dietary habits, especially the increased consumption of animal produce (Falkenmark and Lannerstad 2005; Liu and Savenije 2008). Assuming business as usual, which does not include a substantial production of modern biomass-for-energy, it has been suggested that shortages of fresh water may well become a fact of life for up to 2.5–6.5 billion people by 2050 (World Water Council 2000; Wallace 2002). For instance, water requirements for food consumption are expected to increase greatly in rapidly industrializing countries such as China and India (Falkenmark and Lannerstad 2005; Liu and Savenije 2008). The latter countries are expected to rely increasingly on food export because of limited water availability (Falkenmark and Lannerstad 2005; Liu and Savenije 2008), and this makes it unlikely that they are suitable for large-scale biofuel production.

Still, in case of major net importers of mineral oil, it may be argued that the availability of biofuels on the world market decreases their reliance on the limited number of countries that are suppliers of mineral oil, and that this diversification may contribute to increased overall energy security. Moreover, one would expect that a substantial production of transport biofuel may have a downward effect on mineral oil prices (Eickhout et al. 2008).

On the other hand, there is the matter of the long-term strategy regarding energy security in transport. Transport biofuels are interesting because they may be used as 'drop ins' without a major change in transport technology. But from a long-term perspective, one may argue that major advances should come out of the twin development of higher energy efficiency in transport (Eaves and Eaves 2007; Royal Society 2008) and supply technologies that are much more efficient than photosynthetic organisms in converting solar radiation into usable energy. Using solar cells or concentrated solar power (CSP) to produce $H_2$ for fuel cells or electricity for batteries is an interesting example of the latter (Armor 2005; Ros et al. 2009). Though physical conversion technologies such as solar cells presently have higher costs (excluding external costs) than biofuels, it has been argued that in the long run, it may be better to focus on such physical conversion technologies than taking the 'detour' of biofuels (cf. Lee and Lee 2008).

The alternative of the twin development of higher energy efficiency and more efficient solar energy conversion technologies will also come up in the context of the next three sections (6.4–6.6) which deal with problems linked to the large areas needed for the production of large amounts of biofuels.

## 6.4 Transport Biofuels, Food Prices and Food Security

As to the effect of biofuels on food security, it has already been noted in Chap. 1 that substantial production of transport biofuels will, under market conditions, have an upward effect on food prices. Prices of food crops which also serve as major feedstocks for biofuels are likely to show linkage with fuel prices. The price of sugar in Brazil is now linked to the price of ethanol, and large-scale use of carbohydrates and vegetable oils as transport biofuel feedstocks may be expected to link the prices thereof to fossil fuel prices, corrected for differences in 'energy content' (Naylor et al. 2007; von Braun 2007; Eickhout et al. 2008; Westhoff 2008). The effect of an expanding transport biofuel production on food prices may lead to an increased insufficiency of food for the world's poorest people that are not net food producers and currently spend 50–80% of their total household income on food (Naylor et al. 2007; Runge and Senauer 2007, Daschle et al. 2007; von Braun 2007). Fast expansion of transport biofuel feedstock production might be expected to have a relatively strong upward effect on food prices (von Braun 2007).

The upward effect of transport biofuel production on food prices partially follows from competition between food crops and biofuel crops for good-quality land. This competition occurs both when transport biofuels are based on feedstocks that can be used for food or feed and in the case that feedstocks for lignocellulosic biofuels are grown (Christersson 2008). Thus, the competition extends to part of the lignocellulosic transport biofuels, including biofuels made from lignocellulosic crops, such as *Miscanthus* (e.g. Sørensen et al. 2008), and biofuels from lignocellulosic by-products which are currently used as animal feed (e.g. Linde et al. 2008; Murphy and Power 2008).

However, it may be expected that when lignocellulosic biofuels contribute substantially to transport biofuel production, the upward effect on food prices will be reduced. This is even more so when lignocellulosic crops are converted into electricity for electric traction, because the seed-to-wheel solar energy conversion efficiency is relatively high (see Chap. 2). There will also be an effect when lignocellulosic biomass is converted into biofuel for internal combustion engines, because in this case, more cropped biomass can be turned into transport biofuel. The magnitude of this effect is uncertain, however, as it is not clear how much lignocellulosic biomass can be diverted to transport biofuel production without having a negative impact on soil organic matter levels and animal feed supplies.

The competition between food crops and biofuel crops also appears to apply to biofuel crops which are well adapted to growth on poor-quality land. One example thereof is *Eucalyptus*. Around 1900, *Eucalyptus* was promoted for growth on 'waste lands, where few other trees would grow' (Doughty 2000). However, now it is often grown on good-quality land in competition with food crops, which has led to restrictions on *Eucalyptus* cultivation in some countries (see Chap. 3). More recently, the oil crop *Jatropha* has been promoted because of its ability to grow on marginal land (Kaushik et al. 2007; Achten et al. 2009). However, as evidenced by the eviction of small-scale farmers in Tanzania for large-scale *Jatropha* cropping (Gross 2008) and the replacement of rice production by *Jatropha* in Burma (Ethnic Community Development Forum 2008), in practice, *Jatropha* cultivation may well compete with food production. It has also been found on the basis of experience with *Jatropha* cultivation in Belize, Nicaragua and India that to be competitive, cultivation has to be intensified beyond that of a rain-fed, low-input and drought-resistant crop (Euler and Gorriz 2004). This should be no surprise as on the biodiesel market, *Jatropha* oil also has to compete with vegetable oils, which have been grown under good conditions which are conducive to high yields.

To the extent that the competition between food and transport biofuel crops for good-quality land has been studied for the United States, a rather general upward effect on food prices has been found (Walsh et al. 2003; Johansson and Azar 2007; Schneider et al. 2007). However, there may also be differential effects of biofuel crops on the prices of specific foods. These depend on actual crops that are used for the production of transport biofuels. Elobeid and Hart (2007) have modelled the effect of expanding bioethanol from corn production in the USA and found the biggest impact on food-basket costs in sub-Saharan Africa and Latin America, where corn is a major food grain. A lower impact was expected in Southeast Asia where rice is a major food grain, with countries where wheat and/or sorghum are major staples falling in between. To lessen the effects of biofuel feedstocks on Chinese food prices, in 2008, China began to import cassava as feedstock from Malaysia, the Philippines, Indonesia and Nigeria (Tenenbaum 2008). When China is to heavily rely on cassava as a feedstock for bioethanol production, it would seem likely that prices of this 'poor man's food' may be much increased (Naylor et al. 2007).

Is there a strategy for developing transport biofuels that will not have an upward impact on food prices? The answer to this question should take account of a downward pressure on crop production associated with climate change and increasing land claims associated with agricultural production for an increasing world population with consumption patterns that increasingly favour animal produce (Tilman et al. 2001; Reijnders and Soret 2003; Swedish Environmental Advisory Council 2007; Koneswaran and Nierenberg 2008; von Braun 2007). The latter development will in all probability intensify competition for good-quality land.

To the extent that one relies on crops for the supply of transport biofuel feedstock, while relying on market forces, direct competition with food and feed production therefore seems inevitable, as does an upward effect on food prices. In this respect, there are likely to be quantitative differences linked to the relative yield of transport biofuels per hectare. These differences can be substantial, as shown in Chap. 2. From the data presented in Chap. 2, it would seem that crops proposed for the generation of lignocellulosic feedstock are not necessarily superior to current food crops such as sugar beet and sugar cane as to their net efficiency in converting solar radiation into biomass. How they will perform in net yield of biofuels is rather uncertain because technologies for the conversion of lignocellulosic biomass into transport biofuels are under development, and there is uncertainty about yields that may be possible in the future and the extent to which aboveground biomass should be returned to soils to maintain soil organic matter levels.

Still, it is to be expected that some biofuels would not lead to an upward movement of food prices. Firstly, biofuels produced from what are currently 'wastes', such as organic urban wastes, biomass from forest remediation and residues from forestry and agriculture, which are not used as animal feed, may partly qualify as such. The worldwide amount of these wastes is currently estimated at between 50 and 100 EJ (Swedish Environmental Advisory Council 2007; Lysen and van Egmond 2008). Unfortunately, it is not clear how much thereof is necessary for maintaining soil organic carbon stocks in a steady state to safeguard the future productivity of arable lands and forests (see Chap. 3). However, even when only 10–20% thereof could be diverted to transport biofuel production, this would still represent a substantial contribution to the transport fuel supply.

Another option that has been suggested in this context is growing microalgae (Chisti 2007, 2008; Dismukes et al. 2008; Groom et al. 2008). However, as pointed out in Chap. 2, an overall positive energy conversion efficiency of microalgal biofuels currently seems uncertain. There is also the water demand associated with growing algae, which may, per kilogram of dry weight biomass, be larger than, for example, sugar cane (see Chap. 3). This may lead to claims which may easily compete with agricultural land use.

Still another option is the use of abandoned cropland and lands that sequester little carbon today (Searchinger et al. 2008). The use of terrestrial plants to reclaim deserts may make it possible to harvest lignocellulosic biomass or oil (from, e.g. *Jatropha*). Of course, sustainable productivity of reclaimed drylands is relatively low. Apart from human intervention, actual productivity depends on rainfall (Webb et al. 1978). In tropical and subtropical areas with precipitation below 500 mm $year^{-1}$, aboveground C sequestration may be roughly between 0.15 and 1.5 Mg $ha^{-1}$ $year^{-1}$ (Hadley and Szarek 1981). Increased sustainable yields may be possible by efficient water management and conservation practices (Thomas 2008). In semi-arid (500–750 mm rainfall $year^{-1}$) and sub-humid (750–1,000 mm rainfall $year^{-1}$) environments with relatively high insolation, aboveground C sequestration may amount to 2–3 Mg C $ha^{-1}$ $year^{-1}$ (Lal 2001). In humid Icelandic deserts, restoration activities have led to the sequestration of 0.6–1.1 Mg C $ha^{-1}$ $year^{-1}$ (Ágústdóttir 2004). After reclaiming lands with little C sequestration, there are often competing uses. For example, biomass may be exploited for grazing (Brown 2003; Darkoh 2003; McNeely 2003; Lal 2008; Ludwig et al. 2008), and this may lead to limitations on use for biofuel production.

If compared with reclaimed deserts, biomass production may be higher on currently abandoned and fallow agricultural lands, with proper use of organic amendments and fertilizers and appropriately adopted plant species (Lal and Bruce 1999). Field et al. (2008) and Campbell et al. (2008) have estimated that such lands comprise about $385–472 \times 10^6$ ha. In Chap. 3, it has been estimated that, after restoration of soil organic matter and nutrient levels, the worldwide sustainable feedstock production on such lands may be in the order of 23–28 EJ. When one assumes that the conversion efficiency thereof to transport biofuels is 40–50%, this would allow for the production of 8.6–14 EJ of transport fuels, which would be a substantial contribution to the 85–90 EJ of transport fuels that is currently used in means of transport.

Higher yields may be possible by intensifying cultivation of lands which currently sequester little C and agricultural lands that have been abandoned or fallow, but it is highly doubtful that the biofuels generated in this way could be considered sustainable. As pointed out in Chap. 5, to be successful, this way to exploit fallow and abandoned agricultural land should be viewed by the local population as being in line with their pressing needs. Moreover, for the large-scale cultivation on abandoned and fallow croplands and lands that currently sequester little C, one has to go beyond the market mechanism. For instance, in the case of palm oil, in Malaysia, planting oil palms on abandoned land is currently rare, because degraded land does not provide revenue from initial timber extraction and entails relatively high establishment costs and possibly reduced yields (Wicke et al. 2009). More in general, the profitability of abandoned cropland and land that currently sequesters little C tends to be less than for good-quality agricultural land (Huston and Marland 2003; Johansson and Azar 2007). Therefore, large-scale cultivation of crops for transport biofuels on degraded cropland and in the context of desert reclamation will probably depend on government interventions (see Sect. 6.7).

## 6.5 Is Expanding Biofuel Production a Good Way to Tackle Climate Change?

Searchinger et al. (2008) have shown by careful modelling of land use change that a large and fast expansion of bioethanol production from corn in the USA is counterproductive in tackling climate change. This is largely related to the land use change necessary for displaced food and feed production, which leads to a large desequestration of C. It is likely that a similar conclusion applies to biofuel production in Europe. Regarding Brazil, it has been shown that soybean-based biodiesel is counterproductive in tackling climate change (Reijnders and Huijbregts 2008a), and the same holds for palm-oil-based transport biofuels for which tropical forests are cleared (Danielsen et al. 2008; Fargione et al. 2008; Reijnders and Huijbregts 2008a).

As explained in Chap. 3 and as also pointed out by Fargione et al. (2008), any strategy aiming at expanding transport biofuel production by converting native wooded ecosystems to cropland is likely to be counterproductive in tackling climate change in the coming decades. When abandoned or fallow land over a number of decades has developed into a secondary forest (e.g. Grau et al. 2003), the same will probably apply. In the case that abandoned or fallow agricultural land is peat land, further C losses from soil will be greater than C gains to be made by reductions in fossil fuel use associated with biofuel use, possibly for many centuries (Reijnders and Huijbregts 2007; Danielsen et al. 2008; Fargione et al. 2008; Reijnders and

Huijbregts 2008a). Moreover, restoration of peat bogs may lead to the additional sequestration of C (Tuittila et al. 1999; Dukes 2003).

This leaves a limited number of terrestrial transport biofuel options conducive to tackling climate change. It would seem that from the point of view of effectiveness in limiting climate change, the burning of biofuels in power plants for use in electric traction is to be preferred, as the solar energy conversion efficiency is relatively high and net greenhouse gas emissions relatively low, as indicated in Chaps. 2 and 4.

The biofuel options that can help in tackling climate change are the following:

1. Linking biofuel use to the man-made sequestration of C in soils. This may be done as soil organic matter (Blaine Metting et al. 2001; Read 2008), as 'biochar' (Lehman et al. 2006) or as $CO_2$. Examples of the latter option are: burning biofuels in power plants for electrical traction and sequestration of $CO_2$ in depleted oil and gas fields or aquifers (Haszeldine 2006; Mathews 2008). Similarly, $CO_2$ generated during fermentation or anaerobic conversion of biomass (into compounds such as ethanol, methane and hydrogen) might be captured or sequestered.
2. Limited use of residues of forestry and agriculture as feedstocks for biofuel production. Limitations in part stem from sustainability requirements. The scope of residue removal is, as pointed out in Chap. 3, limited by the need to maintain soil carbon and nutrient stocks, because otherwise future productivity will be impaired. Moreover, some conversions of agricultural residues do not appear to make energetic sense, such as, for instance, the conversion of swine and bovine manure into methane in NW Europe (cf. Chap. 2).
3. Producing biofuels in areas with currently relatively little aboveground biomass, as also discussed in Sect. 6.3, for instance within the framework of reclaiming deserts and reclamation of saline soils and on abandoned lands (Lal and Bruce 1999; Banerjee et al. 2006; Germer and Sauerborn 2007; Danielsen et al. 2008; Lal 2008).

However, from the point of view of mitigating climate change, alternative uses of land with currently little aboveground biomass merit consideration. In the case of abandoned and fallow agricultural mineral soils which may support secondary forests, transport biofuel production – as, for instance, proposed by Cunha da Costa (2004) for deforested areas in the Brazilian Amazon – may be compared to C sequestration linked to re-growth of forest. For instance, in the case of *Saccharum spontaneum* grasslands in Panama that preclude forest regeneration, it has been found that low-cost management options exist for restoring forest cover (Hooper et al. 2005). And also in other contexts, feasible ways have been developed to convert degraded agricultural land into secondary forests (Vieira and Scariot 2006; Cummings et al. 2007). In doing so, high levels of C sequestration may be achieved. Steininger (2000) found in the Brazilian Amazon a biomass accumulation of, on average, $9.1\,\mathrm{Mg\,ha^{-1}\,year^{-1}}$ over a 12-year period, whereas Zarin et al. (2001), studying re-growth of Amazonian forests, found for a 20-year period an average sequestration of about 6 Mg biomass $\mathrm{ha^{-1}\,year^{-1}}$. These values are well above the amount of C that can be displaced by, for example, growing soybeans for biodiesel production (Reijnders and Huijbregts 2008b). Righelato and Spracklen (2007) estimated that as to tackling climate change during the coming decades, the gains per hectare of reforestation would be higher than those from most biofuels. Expansion of secondary forests is now a substantial development in Latin America, Africa and Southeast Asia (Lambin et al. 2003) and is known to have been successful in countries like Puerto Rico, Bhutan, Vietnam, Gambia and Cuba (Chazdon 2008). All in all, an estimated $96 \times 10^6$ ha of abandoned agricultural land has been reforested (Field et al. 2008).

Another alternative that merits consideration in cases where forests may be re-established is agroforestry, which has been advocated as more in line with the pressing needs of local populations than plantations (De Foresta and Michon 1996), and has now been successfully applied in a number of countries, including the Philippines, Peru, Indonesia, China and Vietnam (Chazdon 2008). Agroforestry allows for substantial sequestration of C. For instance, the cacao agroforests of humid Africa sequester up to about 62% of the C of primary forests (Duguma et al. 2001), and rates of C sequestration in agroforests varying from 2–9 Mg C $ha^{-1}$ $year^{-1}$ have been found (Pandey 2002). Of course, comparisons with alternatives such as secondary forests and agroforestry are only meaningful when there is not a fixed mandate for biofuel production. When there is such a fixed mandate, growth of crops for transport biofuel production will move elsewhere, and may, for instance, be associated with cutting down virgin forests (a phenomenon called leakage, discussed in Sect. 1.4). Also, as stated in Sect. 1.4, one should realize that expansion of secondary forests and agroforestry as such do not provide fuels.

## 6.6 Transport Biofuel Production and Nature Conservation

By the beginning of 2008, the share of transport biofuels in the worldwide consumption of transport fuels was below 1%, and land use for transport biofuels was estimated at 13.8 Mha in the USA, Brazil, China and the EU (Renewable Fuels Agency 2008), but significant upward impacts on the conversion of nature into cropland could be noted (OECD-FAO 2007; Nepstad et al. 2008; Chap. 5). Current policy targets for the expansion of transport biofuel production have been estimated to require between 55 and 166 Mha land (Renewable Fuels Agency 2008). Such a major further expansion of transport biofuel production will (ceteris paribus) stimulate the conversion of nature into cropland (Searchinger et al. 2008; Sukhdev 2008).

Moreover, as indicated in Chaps. 2 and 5, it is to be expected that part of a further expansion of transport biofuel production will come from intensification of agriculture. Intensification of agriculture is expected to be associated with higher inputs of nutrients and pesticides and increased irrigation and drainage (Tilman et al. 2001; Datta et al. 2004; Tscharntke et al. 2005; Liira et al. 2008). This, in turn, is expected to lead to a decline in biodiversity among many taxa and a loss of ecosystem services (Tscharntke et al. 2005; Liira et al. 2008).

When, unlike current practice, low-quality land is used for the expansion of biofuel production (as suggested in Sects. 6.3 and 6.5) and there is a fixed amount of transport biofuel to be produced, such as mandated under several current regulations, then – in view of the probably relatively low productivity of the land – larger areas will be needed than in the case of the use of good-quality land. Moreover, as pointed out in Chap. 5, abandoned and fallow lands by themselves rather often harbour substantial biodiversity. When such relatively biodiverse abandoned and fallow agricultural lands are exploited, the negative impact on biodiversity may be large (Huston and Marland 2003; Marland and Obersteiner 2008). On the other hand, as pointed out in Chap. 5, there are also fallow and abandoned agricultural lands with relatively low biodiversity, such as parts of the *Imperata cylindrica* and *Saccharum spontaneum* grasslands, which may be exploited for transport biofuel production with a relatively low impact on biodiversity.

The size of the impact of expanding transport biofuel production is also dependent on other factors. Effects on biodiversity would, for example, be relatively large when current hotspots of biodiversity, such as tropical rainforests, the Cerrado savannah or nature in the Cape region of South Africa (Darkoh 2003; Koh 2007; Danielsen et al. 2008), are converted into land for the production of transport biofuel feedstocks. Also, if precision agriculture and water-efficient irrigation techniques are used for the expansion of feedstock production (cf. Sects. 3.4 and 4.6), the impact thereof on biodiversity may well be lower than in the case of conventional practices, because water consumption and the emissions of nutrients and pesticides may be lower.

Though the impact of a major expansion of transport biofuel production on living nature may be variable, there would seem to be no scope for a major expansion of biofuel production in such a way that biodiversity loss will be zero. It is likely that a major expansion of transport biofuel production will have a major negative effect on biodiversity and ecosystem services.

## 6.7 How to Use Natural Resources for Biofuel Production in a Sustainable Way?

In this book, 'sustainable' is taken to mean that a practice can be continued indefinitely. As explained in Chaps. 2 and 3, this severely limits the extent to which geochemically scarce resources which have been formed in slow geological processes, such as fossil fuels and phosphate ore, can be converted into wastes. Sustainability also requires that renewable resources such as fertile soil, soil organic matter, groundwater and nutrients are maintained and retain their quality. Regarding cropping, this in turn leads to preferences for conservation tillage, much improved nutrient recycling and improved water efficiency. In forestry, this leads to a preference for long rotations and nutrient recycling. Sustainability also limits achievable biomass production. In Chap. 3, biomass production from currently abandoned, including fallow, agricultural land was estimated to have a sustainable yearly yield of approximately 23–28 EJ, about one order of magnitude below the yield suggested by de Vries et al. (2007).

## 6.8 What Government Policy Should One Aim at for Transport Biofuels?

As pointed out in Chap. 1, much of the impetus for the development of transport biofuel production has come from governments. Looking back on the results of government intervention, as discussed in the previous pages of this book, this has been a very mixed blessing. So, are there suggestions for government policy which may be conducive to more beneficial results?

### *6.8.1 Hard Choices*

From the previous sections, it has become clear that in much expanding transport biofuel production, there may often be hard choices to make. Growing crops such as oil palm and corn for transport biofuels may contribute to energy security and mitigate price rises of fossil fuels but may generate for decades more greenhouse gases than fossil transport fuels and have an upward effect on food prices. Growing feedstocks on currently fallow land in the USA and Europe may mitigate the effect of biofuel production on food prices and limit greenhouse gas emissions linked to the transport biofuel life cycle but is probably bad for biodiversity. Intensifying cropping for biofuel feedstocks may be conducive to limiting price increases for food and fuel but may also be unsustainable and have a negative impact on biodiversity (Sukhdev 2008). As large-scale production of transport biofuels may come at significant costs, one may well wonder whether preference should be given to other ways of providing for transport services.

So, we return to the alternative option of better energy efficiency in transport and better solar energy conversion in transport energy supply, raised in Sect. 6.3. Improved energy efficiency is not necessarily an easy alternative. Indeed, in many countries, as to cars, fossil fuel input per person-kilometre has remained virtually constant since the 1973 ‘oil crisis’. Potential gains in energy efficiency due to technical progress were ‘eaten away’ by developments such as preferences for improved comfort and safety, lower occupancy of cars and increased congestion (Schipper et al. 1992). However, it is also known that increased fuel prices are conducive to increased energy efficiency in producing transport fuels and increased energy efficiency of transport (Schipper et al. 1992; Graham and Glaister 2005), and future prices may well be high (GAO 2007), so that may help in making improvements in energy efficiency more successful than was feasible in the past. To the extent that priority is to be given to renewables, the obvious alternative from the point of view of conversion efficiency (as pointed out in Chap. 2) is the use of physical conversion technologies producing electricity, which can either be used for storage in batteries or be converted into $H_2$ for use in fuel cells (Armor 2005; Evans 2008; Ros et al. 2009). Again, this is not an easy alternative. The costs thereof are as yet high, though they are expected to be much reduced over the coming decades (Martinot 2006; Braun 2008).

### *6.8.2 Evident Policy Priorities Regarding the Production of Biofuels*

When biomass is used for powering transport, there are benefits to burning this in power plants for electric traction. Upward impact on food prices and negative impacts on biodiversity thereof will be lower per unit of energy output than in the case of the conversion of biomass to transport biofuels for internal combustion engines. Also, net greenhouse gas emissions may be relatively low. As pointed out in Sect. 1.6, use of this option depends on a much-increased social acceptability of electric traction in car transport. As pointed out in the same section, opinions on the ease with which such increased acceptability may be achieved vary greatly. If the moderate pessimists are right in this respect, government intervention to create incentives for electric cars may be useful.

Problems with the impacts of biofuel production on food prices may furthermore be considerably reduced when biofuel production can be restricted to currently abandoned and fallow agricultural lands and land that currently sequesters little C. As pointed out in Sect. 6.4, this probably cannot be achieved without major government intervention. In the case of drylands, which currently sequester little C, large government investments may be needed in improved water management and conservation practices (Thomas 2008). Such investments are also conducive to an increased resilience against climate change (Thomas 2008). In the case of abandoned agricultural land, incentives should be given that compensate for the financial disadvantages if compared with the cultivation of good agricultural land. Government intervention may focus on limiting the cultivation of biofuel crops to marginal and abandoned agricultural lands or go one step further and establish government-owned companies that grow such crops. Such intervention is not a mission impossible. For instance, the government of Taiwan has focused its support for biodiesel on feedstocks from polluted and fallow agricultural land (Huang and Wu 2008).

Limited use of organic wastes may also be useful in limiting negative side effects of transport biofuel use. In view of current prices for such biofuels (see Chap. 1), government intervention will often be needed to stimulate the production thereof. An evident priority is the improvement of technologies for the conversion of wastes into transport biofuels. As pointed out in Chap. 1, such technologies seem to offer much scope for lowering costs and improving conversion efficiencies. A mandate such as in the Energy Independence and Security Act 2007 of the USA for phase-in of lignocellulosic ethanol may also be helpful. Furthermore, to establish an environmental benefit, criteria have to be established which restrict marketing to biofuels that at least have a predefined benefit. Both the EU and the USA have restrictions in place for the life cycle greenhouse gas emissions of transport biofuels. It would seem useful to extend such criteria, for instance to safeguard soil organic matter and nutrient stocks and to limit negative effects on biodiversity.

## References

Achten WMJ, Verschot L, Franken VJ, Mathijs E, Singh VP, Aerts R, Muys B (2009) Jatropha biodiesel production and use. Biomass Bioenerg in press

Ágústdóttir AM (2004) Revegetation of eroded land and possibilities of carbon sequestration in Iceland. Nutr Cycl Agroecosys 70:241–247

Antoni D, Zverlov VV, Schwarz WH (2007) Biofuels from microbes. Appl Microbiol Biotechnol 77:23–35

Arifeen N, Wang R, Kookos I, Webb C, Koutinas AA (2007) Optimization and cost estimation of novel wheat biorefining for continuous production of fermentation feedstock. Biotechnol Progr 23:872–880

Armor JN (2005) Catalysis and the hydrogen economy. Catal Lett 101:131–135

Bagajewicz M, Sujo D, Martinez D, Savelski M (2007) Driving without petroleum? A comparative guide to biofuels, gas-to-liquids and coal-to-liquids as fuels for transportation. Energy Charter Secretariat, Brussels

Banerjee MJ, Gerhart VJ, Glenn EP (2006) Native plant regeneration on abandoned desert farmland: effects of irrigation, soil preparation, and amendments on seedling establishment. Restor Ecol 14:339–348

Bideaux C, Alfenore S, Cameleyre X, Molina-Jouve C, Uribelarrea J, Guillouet SE (2006) Minimization of glycerol production during the high-performance fed-batch ethanolic fermentation process in *Saccharomyces cerevisiae*, using a metabolic model as a prediction tool. Appl Exp Microbiol 72:2134–2140

Blaine Metting F, Smith JL, Amthor JS, Izaurralde RC (2001) Science needs and new technology for increasing soil carbon sequestration. Climatic Change 51:11–34

Braun AE (2008) Photovoltaics: grid competitive in five years. Semiconductor International March:17–18

Brown G (2003) Factors maintaining plant diversity in degraded areas of northern Kuwait. J Arid Environ 54:183–194

Campbell JE, Lobell DB, Genova RC, Field CB (2008) The global potential of bioenergy on abandoned agriculture lands. Environ Sci Technol 42:5791–5794

Chazdon RL (2008) Beyond deforestation: restoring forests and ecosystem services on degraded lands. Science 320:1458–1460

Chew TL, Bhatia S (2008) Catalytic processes towards the production of biofuels in a palm oil and oil palm biomass-based biorefinery. Bioresour Technol 99:7911–7922

Chisti Y (2007) Biodiesel from microalgae. Biotechnol Adv 25:294–306

Chisti Y (2008) Biodiesel from microalgae beats bioethanol. Trends Biotechnol 26:126–131

Christersson L (2008) Poplar plantations for paper and energy in the south of Sweden. Biomass Bioenerg 32:997–1000

Cummings J, Reid N, Davies I, Grant C (2007) Experimental manipulation of restoration barriers in abandoned eucalypt plantations. Restor Ecol 15:156–167

Cunha da Costa R (2004) Potential for producing bio-fuel in the Amazon deforested areas. Biomass Bioenerg 26:405–415

Danielsen F, Beukema H, Burgess ND, Parish F, Brühl C, Donald PF, Murdiyarsno D, Phalan B, Reijnders L, Struebig M, Fitzherbert EM (2008) Biofuel plantation on forested land: double jeopardy for biodiversity and climate. Conserv Biol in press

Darkoh MBK (2003) Regional perspectives on agriculture and biodiversity in the drylands of Africa. J Arid Environ 54:261–279

Daschle T, Runge CF, Senauer B (2007) Debating the tradeoffs of corn-based ethanol: myth versus reality. Foreign Affairs September/October

Datta KK, Tewari L, Joshi PK (2004) Impact of subsurface drainage on improvement of crop production and farm income in north-west India. Irrig Drain Syst 18:43–56

De Foresta H, Michon G (1996) The agroforest alternative to *Imperata* grasslands: when smallholder agriculture and forestry reach sustainability. Agroforest Syst 36:105–120

de Vries BJM, van Vuuren DP, Hoogwijk MM (2007) Renewable energy sources: their global potential for the first-half of the 21st century at a global level: an integrated approach. Energ Policy 35:2590–2610

Dismukes GC, Carrieri D, Bennette N, Ananyev GM, Posewitz MC (2008) Aquatic phototrophs: efficient alternatives to land-based crops for biofuels. Curr Opin Biotechnol 19:235–240

Doughty RW (2000) The eucalyptus: a natural and commercial history of the gum tree. Johns Hopkins University, Baltimore

Duguma B, Gockowski J, Bakala J (2001) Smallholder cacao (*Theobroma cacao* Linn.) cultivation in agroforestry systems of west and central Africa: challenges and opportunities. Agroforest Syst 51:177–188

Dukes JS (2003) Burning buried sunshine: human consumption of ancient solar energy. Climatic Change 61:31–44

Dürre P (2008) Fermentative butanol production: bulk chemical and biofuel. Ann NY Acad Sci 1125:353–362

Eaves J, Eaves S (2007) Renewable corn-ethanol and energy security. Energ Policy 35:5958–5963

Eickhout B, van den Born GJ, Notenboom J, van Oorschot M, Ros JPM, van Vuuren DP, Westhoek HJ (2008) Local and global consequences of the EU renewable directive for biofuels. Milieu en Natuur Planbureau Bilthoven. http://www.mnp.nl

Elobeid A, Hart C (2007) Ethanol expansion in the food versus fuel debate: how will developing countries fare? J Agric Food Ind Organ 5(2):article 6. http://www.bepress.com/jafio/vol5/iss2/art6/

Ethnic Community Development Forum (2008) Biofuel by decree: unmasking Burma's bio-energy fiasco. http://www.terraper.org/file_upload/BiofuelbyDecree.pdf

Euler H, Gorriz FD (2004) Case study *Jatropha curcas*. Global Facilitation Unit for Underutilized Species and Deutsche Gesellschaft für Technische Zusammenarbeit, Frankfurt

Evans RL (2008) Energy conversion chain analysis of sustainable energy systems: a transportation case study. B Sci Technol Soc 28:128–137

Falkenmark M, Lannerstad M (2005) Consumptive water use to feed humanity – curing a blind spot. Hydrol Earth Syst Sci 9:15–28

Fargione J, Hill J, Tilman D, Polasky S, Hawthorne P (2008) Land clearing and the biofuel carbon debt. Science 319:1235–1238

Field CB, Campbell JE, Lobell DB (2008) Biomass energy: the scale of the potential resource. Trends Ecol Evol 23:65–72

GAO (United States Government Accountability Office) (2007) Crude oil: uncertainty about future oil supply makes it important to develop a strategy for addressing a peak and decline in oil production. GAO-07-283. http://www.gao.gov/new.items/d07283.pdf

Germer J, Sauerborn J (2007) Estimation of the impact of oil palm plantation establishment on greenhouse gas balance. Environ Dev Sust. doi:10.1007/s1066800690801

Graham DJ, Glaister S (2005) Decomposing the determinants of road traffic demand. Appl Econ 37:19–28

Grau HR, Aide TM, Zimmerman JK, Thomlinson JR, Helmer E, Zou X (2003) The ecological consequences of socioeconomic and land-use changes in postagriculture Puerto Rico. BioScience 53:1159–1168

Groom MJ, Gray EM, Townsend PA (2008) Biofuels and biodiversity: principles for creating better policies for biofuel production. Conserv Biol 22:602–609

Gross M (2008) Not in our backyard. Curr Biol 18:R227–R228

Hadley NF, Szarek SR (1981) Productivity of desert ecosystems. BioScience 31:747–753
Haszeldine RS (2006) Deep geological $CO_2$ storage: principles reviewed, and prospecting for bioenergy disposal sites. Mitigation Adaptation Strategies Global Change 11:369–393
Hayes DJ (2008) An examination of biorefining processes, catalysts and challenges. Catal Today in press
Hooper E, Legendre P, Condit R (2005) Barriers to forest regeneration of deforested and abandoned land in Panama. J Appl Ecol 42:1165–1174
Huang H, Ramaswamy S, Tschirner UW, Ramarao BV (2008) A review of separation technologies in current and future biorefineries. Sep Purif Technol 62:1–21
Huang Y, Wu J (2008) Analysis of biodiesel promotion in Taiwan. Renew Sust Energ Rev 12: 1176–1186
Huston MA, Marland G (2003) Carbon management and biodiversity. J Environ Manag 67:77–86
Jepma C (2008) An editorial comment. Re: Read P. Biosphere carbon stock management: addressing the threat of abrupt climate change in the next few decades. Climatic Change 87:343–346
Johansson DJA, Azar C (2007) A scenario based analysis of land competition between food and bioenergy production in the US. Climatic Change 82:267–291
Kadam KL, Chin CY, Brown LW (2008) Flexible biorefinery for producing fermentation sugars, lignin and pulp from corn stover. J Ind Microbiol Biotechnol 35:331–341
Kamm B, Kamm M (2004) Principles of biorefineries. Appl Microbiol Biotechnol 64:137–145
Kaushik N, Kumar K, Kumar S (2007) Potential of *Jatropha curcas* for biofuels. J Biobased Mat Bioenerg 1:301–314
Koh LP (2007) Potential habitat and biodiversity losses from intensified biodiesel feedstock production. Conserv Biol 21:1373–1375
Koneswaran G, Nierenberg D (2008) Global farm animal production and global warming: impacting and mitigating climate change. Environ Health Persp 116:578–582
Lal R (2001) Potential of desertification control to sequester carbon and mitigate the greenhouse effect. Climatic Change 51:35–72
Lal R (2008) Crop residues as soil amendments and feedstock for bioethanol production. Waste Manage 28:747–758
Lal R, Bruce JP (1999) The potential of world cropland soils to sequester C and mitigate the greenhouse effect. Environ Sci Policy 2:177–185
Lambin EF, Geist HJ, Lepers E (2003) Dynamics of land-use and land-cover change in tropical regions. Annu Rev Environ Resour 28:205–241
Lee D, Lee D (2008) Biofuel economy and hydrogen competition. Energy Fuels 22:177–181
Lehmann J, Gaunt J, Rondon M (2006) Bio-char sequestration in terrestrial ecosystems – a review. Mitigation Adaptation Strategies Global Change 11:395–419
Liira J, Schmidt T, Aavik T, Arens P, Augenstein I, Bailey D, Billeter R, Bukáček R, Burel F et al. (2008) Plant functional group competition and large-scale species richness in European agricultural landscapes. J Veg Sci 19:3–14
Linde M, Galbe M, Zacchi G (2008) Bioethanol production from non-starch carbohydrate residues in process streams from a dry-mill ethanol plant. Bioresour Technol 99:6505–6511
Liu J, Savenije HHG (2008) Food consumption patterns and their effect on water requirement in China. Hydrol Earth Syst Sci 12:887–898
Ludwig F, de Kroon H, Prins HHT (2008) Impacts of savanna trees on forage quality for large African herbivore. Oecologia 155:487–496
Lysen E, van Egmond S (eds) (2008) Assessment of biomass potentials and their links to food, water, biodiversity, energy demand and economy. http://www.mnp.nl/en/publications/2008/Assessment_of_global_biomass_potentials_MainReport.html
Macedo IC, Seabra JEA, Silva JEAR (2008) Greenhouse gases emissions in the production and use of ethanol from sugarcane in Brazil: the 2005/2006 averages and a prediction for 2020. Biomass Bioenerg 32:582–595
Marland G, Obersteiner M (2008) Large-scale biomass for energy, with considerations and caution: an editorial comment. Climatic Change 87:335–342

Martinot E (2006) Renewable energy gains momentum: global markets and policies in the spotlight. Environment 48(6):26–43
Mathews JA (2008) Carbon-negative biofuels. Energ Policy 36:940–945
McNeely JA (2003) Biodiversity in arid regions: values and perceptions. J Arid Environ 54:61–70
Murphy JD, Power NM (2008) How can we improve the energy balance of ethanol production from wheat. Fuel 87:1799–1806
Naylor RL, Liska AJ, Burke MB, Falcon WP, Gaskell JC, Rozelle SD, Cassman KG (2007) The ripple effect: biofuels, food security, and the environment. Environment 49(9):30–43
Nepstad DC, Stickler CM, Soares-Filho B, Merry F (2008) Interactions among Amazon land use, forests and climate: prospects for a near-term forest tipping point. Philos T R Soc B 363: 1737–1746
OECD-FAO (2007) OECD-FAO Agricultural Outlook 2007–2016. OECD, Paris
Pandey DN (2002) Carbon sequestration in agroforestry systems. Clim Policy 2:367–377
Read P (2008) Biosphere carbon stock management: addressing the threat of abrupt climate change in the next few decades: an editorial essay. Climatic Change 87:305–320
Reijnders L, Huijbregts MAJ (2007) Life cycle greenhouse gas emissions, fossil fuel demand and solar energy conversion efficiency in European bioethanol production for automotive purposes. J Clean Prod 15:1806–1812
Reijnders L, Huijbregts MAJ (2008a) Palm oil and the emission of carbon-based greenhouse gases. J Clean Prod 16:477–482
Reijnders L, Huijbregts MAJ (2008b) Biogenic greenhouse gas emissions linked to the life cycles of biodiesel derived from European rapeseed and Brazilian soybeans. J Clean Prod 16: 1943–1948
Reijnders L, Soret S (2003) Quantification of the environmental impact of different dietary protein choices. Am J Clin Nutr 78:664S–668S
Renewable Fuels Agency (2008) The Gallagher review of the indirect effects of biofuels. Renewable Fuels Agency, St Leonards-on-Sea (East Sussex, UK)
Righelato R (2008) Forests or fuel? Ecologist February:33
Righelato R, Spracklen DV (2007) Carbon mitigation by biofuels or by saving and restoring forests? Science 317:902
Ros J, Nagelhout D, Montfoort J (2009) New environmental policy for system innovation: Casus alternatives for fossil motor fuels. Appl Energ 86:243–250
Rowlands WN, Masters A, Maschmeyer T (2008) The biorefinery – challenges, opportunities, and an Australian perspective. B Sci Technol Soc 28:149–158
Royal Society (2008) Sustainable biofuels: prospects and challenges. http://royalsociety.org
Runge CF, Senauer B (2007) How biofuels could starve the poor. Foreign Affairs May/June
Schipper L, Meyers S, Howarth RB, Steiner R (1992) Energy efficiency and human activity: past trends, future prospects. Cambridge University, Cambridge
Schneider UA, McCarl BA, Schmid E (2007) Agricultural sector analysis on greenhouse gas mitigation in US agriculture and forestry. Agr Syst 94:128–140
Searchinger T, Heimlich R, Houghton RA, Dong F, Elobeid A, Fabiosa J, Tokgoz S, Hayes D, Yu T (2008) Use of U.S. croplands for biofuels increases greenhouse gases through emissions from land-use change. Science 319:1238–1240
Sørensen A, Teller PJ, Hilstrøm T, Ahring BK (2008) Hydrolysis of *Miscanthus* for bioethanol production using dilute acid presoaking combined with wet explosion pre-treatment and enzymatic treatment. Bioresour Technol 99:6602–6607
Steininger MK (2000) Secondary forest structure and biomass following short and extended land-use in central ands southern Amazonia. J Trop Ecol 16:689–708
Sukhdev P (2008) The economics of ecosystems & biodiversity. European Communities, Brussels
Swedish Environmental Advisory Council (2007) Scenarios on economic growth and resource demand. Statens Offentliga Utreningar, Stockholm
Tenenbaum DJ (2008) Food vs. Fuel: diversion of crops could cause more hunger. Environ Health Persp 116:A254–A257

Thomas RJ (2008) Opportunities to reduce the vulnerability of dryland farmers in Central and West Asia and North Africa to climate change. Agr Ecosyst Environ 126:36–45

Tilman D, Fargione J, Wolff B, D'Antonio C, Dobson A, Howarth R, Schindler D, Schlesinger WH, Simberloff D, Swackhamer D (2001) Forecasting agriculturally driven global environmental change. Science 292:281–284

Tscharntke T, Klein AM, Kruess A, Steffan-Dewenter I, Thies C (2005) Landscape perspectives on agricultural intensification and biodiversity – ecosystem service management. Ecol Lett 8: 857–874

Tuittila E, Komulainen V, Vasander H, Laine J (1999) Restored cut-away peatland as a sink for atmospheric $CO_2$. Oecologia 120:563–574

Tyner WE (2008) The US ethanol and biofuels boom: its origins, current status, and future prospects. BioScience 58:646–653

Vedenov D, Wetzstein M (2008) Toward an optimal U.S. ethanol fuel subsidy. Energ Econ 30:2073–2090

Vieira DLM, Scariot A (2006) Principles of natural regeneration of tropical dry forests for restoration. Restor Ecol 14:11–20

von Braun J (2007) The world food situation. International Food Policy Research Institute, Washington DC

Wallace JS (2002) Increasing agricultural water use efficiency to meet future food production. Agr Ecosyst Environ 82:105–119

Walsh ME, de la Torre Ugarte DG, Shapouri H, Slinsky SP (2003) Bioenergy crop production in the United States: potential quantities, land use changes, and economic impacts on the agricultural sector. Environ Resour Econ 24:313–333

Webb W, Szarek S, Lauenroth W, Kinerson R, Smith M (1978) Primary productivity and water use in native forest, grassland, and desert ecosystems. Ecology 59:1239–1247

Westhoff P (2008) Farm commodity prices: why the boom and what happens now? Choices 23(2):6–10

Wicke B, Dornburg V, Junginger M, Faaij A (2009) Different palm oil production systems for energy purposes and their greenhouse gas implications. Biomass Bioenerg in press

World Water Council (2000) World water vision. Earthscan, London

Yazdani SS, Gonzalez R (2007) Anaerobic fermentation of glycerol: a path to economic viability for the biofuels industry. Curr Opin Biotechnol 18:213–219

Zarin DJ, Ducey MJ, Tucker JM, Salas WA (2001) Potential biomass accumulation in Amazonian regrowth forests. Ecosystems 4:658–668

# Index